GORILLA MOUNTAIN

GORILLA MOUNTAIN

the story of wildlife biologist

AMY VEDDER

by Rene Ebersole

Franklin Watts
A Division of Scholastic Inc.
New York • Toronto • London • Auckland • Sydney
Mexico City • New Delhi • Hong Kong
Danbury, Connecticut

Joseph Henry Press
Washington, D.C.

Author's Acknowledgments

Amy Vedder, you are truly an inspiration. Thank you for captivating me with your wonderful adventures and life lessons. This book would not have been possible without your patience, generosity, and willingness to talk at all hours of the day and night. Thanks also to Bill Weber, Marion Vedder, Marilyn Vedder, Barbara Vedder, and Noah Weber for sharing your memories and stories of Amy. Many of Amy's colleagues at the Wildlife Conservation Society, including George Schaller, John Robinson, and Omari Ilambu, contributed to my education in the challenges facing conservationists who are helping Africa's wildlife persevere despite rampant poaching and horrific wars. A number of other scientists took time out of their busy schedules to review the manuscript. They include primatologist Liz Williamson, anthropologist Colin Groves, and graduate student Corinna Ross. Finally, I'm extremely grateful to my husband, Michael Newhouse, and our families for your endless support and encouragement.—RE

Cover photo: Wildlife biologist Amy Vedder demonstrates the techniques for conducting a field survey of gorillas in Gabon, Africa. She enters data from the survey in a field notebook and uses an altimeter to record the elevation.

Cover design: Michele de la Menardiere

Library of Congress Cataloging-in-Publication Data

Ebersole, Rene, 1974-
Gorilla mountain : the story of wildlife biologist Amy Vedder / Rene Ebersole.
p. cm.— (Women's adventures in science) .
Includes bibliographical references and index.
ISBN 0-531-16779-8 (lib. bdg.) ISBN 0-309-09551-4 (trade pbk.) ISBN 0-531-16954-5 (classroom pbk.)
1. Vedder, Amy—Juvenile literature. 2. Biologists—United States—Juvenile literature.
3. Gorilla—Rwanda. 4. Wildlife conservation—Rwanda. I. Title.

QL31.V37E24 2005
599.884'092—dc22

2005000823

Printed in the United States of America.
1 2 3 4 5 6 7 8 9 10 R 14 13 12 11 10 09 08 07 06 05

About the Series

The stories in the *Women's Adventures in Science* series are about real women and the scientific careers they pursue so passionately. Some of these women knew at a very young age that they wanted to become scientists. Others realized it much later. Some of the scientists described in this series had to overcome major personal or societal obstacles on the way to establishing their careers. Others followed a simpler and more congenial path. Despite their very different backgrounds and life stories, these remarkable women all share one important belief: the work they do is important and it can make the world a better place.

Unlike many other biography series, *Women's Adventures in Science* chronicles the lives of contemporary, working scientists. Each of the women profiled in the series participated in her book's creation by sharing important details about her life, providing personal photographs to help illustrate the story, making family, friends, and colleagues available for interviews, and explaining her scientific specialty in ways that will inform and engage young readers.

This series would not have been possible without the generous assistance of Sara Lee Schupf and the National Academy of Sciences, an individual and an organization united in the belief that the pursuit of science is crucial to our understanding of how the world works and in the recognition that women must play a central role in all areas of science. They hope that *Women's Adventures in Science* will entertain and enlighten readers with stories of intellectually curious girls who became determined and innovative scientists dedicated to the quest for new knowledge. They also hope the stories will inspire young people with talent and energy to consider similar pursuits. The challenges of a scientific career are great but the rewards can be even greater.

Other Books in the Series

Beyond Jupiter: The Story of Planetary Astronomer Heidi Hammel

Bone Detective: The Story of Forensic Anthropologist Diane France

Forecast Earth: The Story of Climate Scientist Inez Fung

Gene Hunter: The Story of Neuropsychologist Nancy Wexler

Nature's Machines: The Story of Biomechanist Mimi Koehl

People Person: The Story of Sociologist Marta Tienda

Robo World: The Story of Robot Designer Cynthia Breazeal

Space Rocks: The Story of Planetary Geologist Adriana Ocampo

Strong Force: The Story of Physicist Shirley Ann Jackson

Contents

Introduction ix

1. Close Encounter 1
2. Journey to the Mountain 5
3. In the Land of a Thousand Hills 21
4. Midnight Voyage among the Volcanoes 29
5. Amy the Ape 37
6. Under Attack 49
7. Close Calls 57
8. Going Beyond Gorillas 65
9. Murder on the Mountain 73
10. Family Life in a Foreign Land 79
11. Guerrilla Wars 91
12. Pablo's Surprise 97

Timeline 104

About the Author 106

Glossary and Metric Conversion Chart 107

Further Resources and Bibliography 109

Index 111

Series Advisory Boards 115

Credits 117

INTRODUCTION

It's a Wild Life

In the late 1970s, the number of mountain gorillas in Africa had dropped dangerously low. So low, in fact, that they were about to disappear altogether. Amy Vedder wanted to save them. Fortunately, she was too young and too naive to think she couldn't do it. Amy is a wildlife biologist—she studies what makes animals tick. She went to Africa in 1978 intent on learning enough about mountain gorillas to keep them from becoming extinct.

Not only did Amy help save the gorillas, she had the adventure of a lifetime. She even became an honorary member of a gorilla family!

Amy also learned a lot about herself and her desire to protect wild animals from danger. Since then she has traveled around the world helping wildlife. One month she might be in Mongolia, learning how to protect endangered gazelles. The next month might find her closer to home, observing recovered wolves in Wyoming's Yellowstone National Park. Back in New York, she helps city kids visiting the Bronx Zoo learn about threatened wildlife and the need for conservation.

Thanks to Amy's hard work, many animal populations have grown stronger. Their habitats have become protected parks—and that makes Amy feel good.

How did Amy Vedder get into the business of saving wildlife? Simple: She followed her instincts, and they led her to science.

How did Amy get into the business of saving ***wildlife?***

Simple:

She followed her instincts, and they led her to ***science.***

1

Close Encounter

A bone-chilling rain poured over Amy Vedder as she huddled on the forest floor, surprised by what had just happened. Perched under her arm sat a four-year-old mountain gorilla named Pablo. All his life Pablo had turned to his mother for shelter from stormy weather. Now, abandoned, he seemed to have decided that Amy—the white ape in a raincoat—would be a fitting stand-in mother.

Mountain gorilla Pablo *(above)* was four years old when he tried to "adopt" wildlife biologist Amy Vedder *(opposite)* in Rwanda, Central Africa, in 1978. Amy's mission there was to help Pablo and his fellow mountain gorillas survive.

Just moments earlier Amy had noticed that Pablo was acting strange. He was walking toward her with his chin lowered. This was not his usual "run by and give her a kick" sort of way.

What's he up to? she thought. *Pablo doesn't usually come near unless he's trying to get me to play.*

She stopped scribbling notes and waited for Pablo to give her a clue. He inched closer. Then Pablo tucked his head under Amy's arm and snuggled close.

~ An Unlikely Pair

It was a move she had seen Pablo pull only on his mother. Amy's shelter must have looked meager compared with the mother gorilla's massive limb and its heavy covering of hair. But Pablo didn't seem to mind. Glued to Amy's side, he looked grateful for whatever warmth she could offer. Amy's heart swelled.

She was thrilled by the idea of being accepted by a gorilla. Yet she knew there was no way she could fill in as Pablo's mother. After all, she's a human and he's an ape. Amy knew she couldn't be there for Pablo day and night. Next she considered the impact on the gorillas' lives if she took Pablo under her wing—and the reason she was sitting in the middle of a Central African rain forest in the first place. Amy was trying to crack a mystery: What would it take to keep these mountain gorillas alive?

Crouched with Pablo in the pouring rain, the biologist in Amy reasoned that his mother surely knew what she was doing when she left her four-year-old son behind the day before to join another gorilla group. *Pablo is old enough to fend for himself,* Amy thought. *If he gets into trouble, the other gorillas in his group will probably come to his rescue.*

She was thrilled by the idea of being accepted by a gorilla. Yet she knew there was no way she could fill in as Pablo's mother.

No matter what happened, Amy decided, Pablo simply had to climb out from under her arm and get on with his life as a gorilla. With a brief farewell squeeze, Amy pushed Pablo away.

~ *Dad to the Rescue*

Over the next few days, Amy watched—and worried—as none of the older female gorillas offered Pablo any warmth. With his crossed eyes and crooked grin, the little gorilla approached Amy again and again, only to find her arms clamped uninvitingly at her sides.

Then a surprise: When dark clouds descended and showered rain on the jungle, Amy saw Beethoven, the kinglike male silverback in the group, shelter Pablo beneath his hulking chest. Every night after that, when Beethoven had finished pulling up plant stems, vines, and leaves and shaping them into a warm, dish-shaped nest, he allowed Pablo to make a crude nest of his own nearby.

Mature silverback males—named for the silvery hair that grows on their backsides as they age—reign supreme in a gorilla group.

Mountain gorillas like Beethoven have the perfect form—a sphere covered by shaggy hair—for living in their cold, wet environment. With its arms crossed and its shoulders hunched, a mountain gorilla can wait out a storm for several hours.

They father most of the offspring, but they generally leave child-rearing to the mothers. Pablo was Beethoven's exception.

Amy took great pleasure in watching this bonding between silverback and son. Such a rare event was proof that she had made the right decision to send Pablo packing. Still, she couldn't help wondering what the future held for him. Amy couldn't imagine him as a mature adult, much less a powerful leader like Beethoven. All she could hope was that Pablo would survive.

When summer rolled around,
the ***girls*** *piled into their parents' car*
for a three-month stay

in New York's Adirondack Mountains,
where the family had a cabin.
The Adirondacks were ***kid heaven****.*

2

Journey to the Mountain

Animals and nature have fascinated Amy Vedder since she was a girl wandering through the forests and fields behind her parents' house. Her family lived on the outskirts of Palatine Bridge, a town of about 800 people in upstate New York.

Even at the age of 5, Amy was drawn to the beauty of nature *(opposite)*. Her concern for animals *(above)* increased as she grew older.

Born on March 24, 1951, Amy is—as she likes to say—the "lower middle Vedder" in the family. She is the second youngest of her sisters: Nancy, Barbara, and Marilyn. Together, the Vedder girls enjoyed three seasons in their rural backyard. They kicked leaves and carved pumpkins in the fall, built igloos and tramped through an icy creek in the winter, and probed that same creek for crayfish and snails in the spring.

When summer rolled around, the girls piled into their parents' car for a three-month stay in New York's Adirondack Mountains, where the family had a cabin. The Adirondacks were kid heaven. There was the water, the woods, and lots of kids to play with! The summer revolved around swimming and waterskiing on Caroga Lake, riding bikes on tar roads, playing games on an abandoned clay tennis court, and running through the woods.

However, life on the lake wasn't always perfect. One summer Amy spent several days in the hospital after a waterskiing accident. The 14-year-old had been skiing a few laps around the lake.

Amy shows off her waterskiing skills on Caroga Lake, New York, in 1962.

Her father was steering the boat, heading back toward the family's dock. Amy decided she wanted to ski one more lap around the lake, but before she could tell her dad to keep going, she collided with a wooden raft.

Amy felt like the air had been sucked from her lungs. Frozen with fear, she hung onto the edge of the raft, thinking *Just don't let go*.

Amy's cousin, Peter, had seen the crash from his family's house on the lake. He raced down to the water and jumped in, hoping to reach Amy in time. Pale and scared, Amy felt herself being pulled out of the water and lifted up onto the raft. Later, at the hospital, doctors worked for several hours to repair a hole in her intestine.

After a few weeks of taking it easy, Amy returned to her normal routine. She hung out with her sisters and friends at a place in the woods they called Big Rock. She wandered alone through the forest, peering wide-eyed at red squirrels, chipmunks, and pink flowers called lady's slippers.

Back in Palatine Bridge, Amy spent the rest of the year surrounded by animals. Her family had pets—dogs, parakeets, a rabbit, and fresh-water fish—but other animals were brought to her father as patients.

The Vedder girls hang out outside their summer home in the Adirondacks in 1960. From left, they are Barbara (age 11), Amy (age 9), Marilyn (age 6), and Nancy (age 13).

Chuck Vedder, Amy's father, was a veterinarian. Living in a rural area, Chuck specialized in caring for the cows and horses on nearby farms. He also had a small animal hospital on the Vedder homestead, where he treated cats and dogs.

Amy's mother, Marion, a trained nurse, spent most of her time taking care of Chuck's business and looking after the girls. She also assisted with dog and cat surgeries in the animal hospital. Amy and her sisters lent a hand too, cleaning out kennels and answering the phone. And each Vedder girl witnessed the delivery a calf or a foal.

Alone or with her family's terrier Pip, Amy wandered the woods near her parents' house, looking for rare wildflowers or daydreaming on the banks of a nearby creek.

Because their father worked long hours, the girls didn't see him very much. Often he said goodnight over a two-way radio. Chuck and Marion worked hard so they could send Nancy, Barbara, Amy, and Marilyn to college. Amy's parents had been the first members of their families to go to college, and they knew money spent on education could pave the way to a fulfilling future for their girls. A good education was a priority.

~ "Always New Things to Discover"

All of the Vedder girls attended kindergarten through third grade in a small stone schoolhouse a short walk from their home. When Amy was in third grade, she demonstrated for her mother how well she could read a newspaper—out loud, upside down, and fast. Amy's mother wondered if her third child was gifted or just a little strange. Amy explained, "I'm a helper, Mother. I sit across from someone else in my class and help her learn to read."

Amy, age 13, soaks up the sun in the field behind her home in Palatine Bridge. In school she was a standout in basketball, softball, and cheerleading. Off the field, she excelled at science and math.

From elementary through high school, Amy kept active, both during and after school. She sampled everything from science fairs—where she proudly showed off her large rock and shell collections—to basketball, softball, and cheerleading. At the same time, she studied hard, and it paid off. When Amy graduated from high school in 1969, she was the top student in her class.

As Amy prepared for college, she struggled with what to study—math or biology? Those were her favorite subjects. Choosing between the two was tough, but she settled on the subject she thought would engage her most in the long run—biology. From her summers spent exploring the woods, Amy knew what fun it was to watch animals in their natural environment. There will always be new, unpredictable things to discover, she decided.

~ *Romance Under the Elms*

Amy applied to just one college: Swarthmore, a small Quaker school in Pennsylvania. Not only did Swarthmore have a well-respected biology program, it also had a small campus surrounded by woods. As at home, those woods would provide a place for Amy to get away from it all when she wanted.

Within days of arriving at Swarthmore, Amy met a young man named Bill Weber. Bill was a sophomore lacrosse player and psychology major. He too had grown up in a small upstate New York town and, like Amy, had spent much of his childhood exploring the woods and fields around his house. On the day they met, Bill was sitting on the lawn at Swarthmore when he noticed a pretty freshman girl walking toward him. She had long, straight brown hair and a dazzling smile. It would take Bill a full semester to persuade Amy to go out with him. "She wasn't sure what to make of this very long-haired jock," he recalled.

Amy *(lower left)* anchors a human pyramid with some of her fellow lacrosse players at Swarthmore College.

As in high school, Amy was a focused student, even when the social and political scenes around her were distracting. The 1960s were marked by the Vietnam War and the assassinations of President John F. Kennedy and Martin Luther King, Jr. Protests on college campuses were common, and Swarthmore was a hotbed for peace protests and strikes. In the midst of this, Amy also concentrated on her studies and sports.

Amy's intense focus left her little time to relax. During spring break of her sophomore year, however, she decided to kick back. Amy climbed into a van with Bill and their group of friends and set off for Jonathan Dickinson State Park in southeast Florida.

Returning from a road trip to Florida, Amy *(bottom right)* and her boyfriend, Bill Weber *(atop the van at far left)*, pose with their friends.

Beneath swaying palm trees, they camped, canoed a maze of overgrown waterways, and ate fresh grapefruit until their stomachs hurt.

By the time the winter holidays rolled around, Bill knew he loved and respected Amy so much he wanted to share all of life's adventures with her. Over winter break, Bill drove up to the Vedder house and offered Amy a diamond engagement ring.

Two days before Bill was to graduate in June 1972, Amy and Bill's closest friends and family members gathered on the shady Swarthmore campus. Wearing a simple flowered dress, Amy married the man who had become her best friend. Getting married was the last thing Amy had expected to do at the age of 21. But, she said, "It just felt right."

In June 1972, Amy and Bill prepare to exchange wedding vows in the courtyard of Swarthmore College.

~ The Toughest Job You'll Ever Love

Soon after Amy graduated from college the next year, she and Bill decided to go on a grand adventure. Because they both had learned French in high school, the newlyweds applied for jobs with the U.S. Peace Corps in one of two French-speaking regions, the Pacific Islands and Central Africa. When they learned they'd been selected to join more than 100 volunteer schoolteachers headed for Zaire, they thought, *Great—where's that?* They quickly learned that Zaire—known today as the Democratic Republic of the Congo (DRC)—was in the heart of Central Africa.

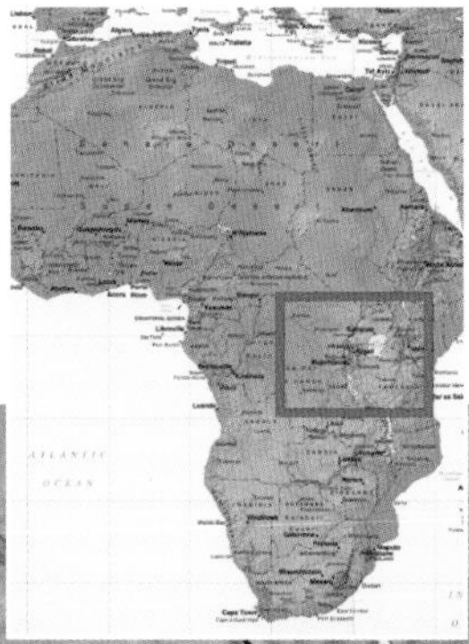

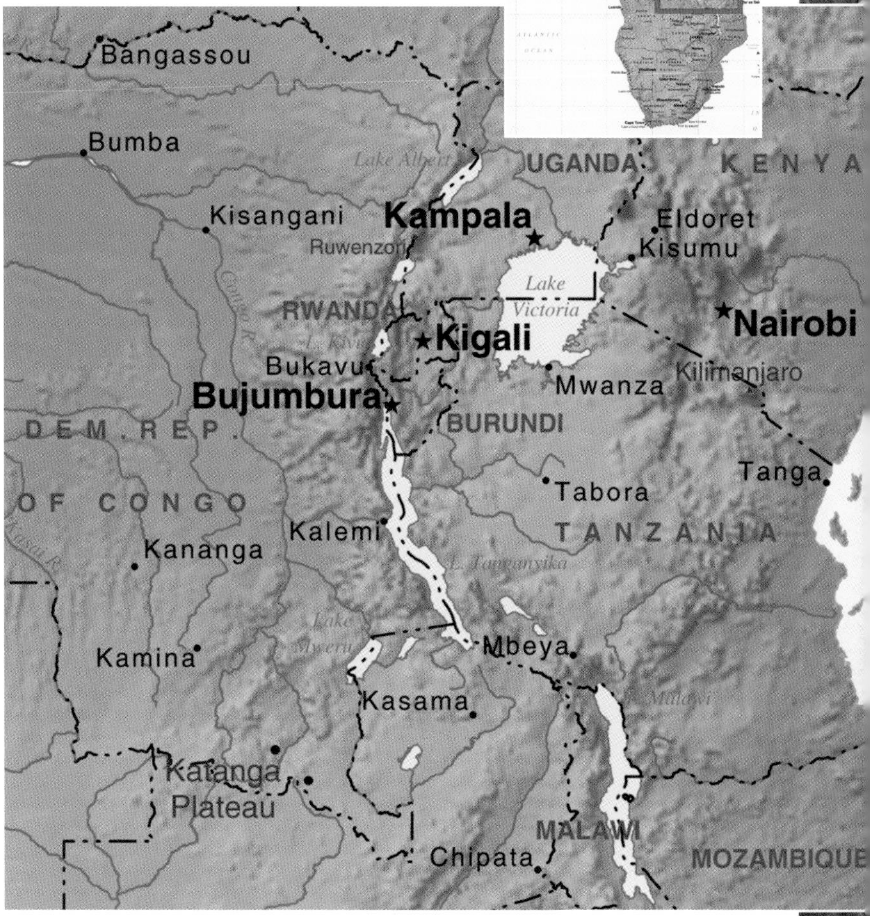

Amy's parents worried about her plan to live in Africa. Wars were being fought there. Political instability plagued several nations in the region where their daughter and new son-in-law were headed. But Amy reassured them that everything would be fine.

On her way to Africa, though, Amy had to wonder if her parents had been right. En route from London to eastern Zaire, the pilot of her plane announced that a scheduled stop in Rwanda was being shifted to neighboring Uganda. They would later learn the reason. A coup, an attempt to overthrow the government, had occurred that day in Rwanda and the airport was closed.

Amy was uncomfortable with the decision to land without warning in Uganda. At the time, Uganda was ruled by an unpredictable and ruthless dictator, Idi Amin. Just two years earlier, in 1971, Amin had ordered the deaths of hundreds of Ugandans. Amy felt nervous about landing in his country, a place with which the United States had broken all ties.

~ *Detained and Searched*

After quickly refueling, the plane took to the air again toward Zaire, and Amy relaxed in her seat. But then there was another announcement from the cockpit: "Ugandan air traffic control has ordered us to return to the airport." As they later found out, Amin had been at the airport in Uganda to meet a visiting foreign president when Amy and Bill's plane took off. Suspicious about who was on the plane, he recalled it to the airport.

For the next 50 hours, Amy, Bill, and the other Peace Corps volunteers were held captive at Entebbe [en-TEB-uh] Airport. Khaki-clad soldiers with AK-47 assault rifles slung over their shoulders stood guard over the group. The 112 volunteers waited—bored but tense—as the Ugandan government searched their bags and interviewed passengers at random.

A few Ugandans who worked at the airport apologized to Amy and her friends. "Please don't judge our country by this experience," they said. Amy knew the situation was a mark of the Ugandan government, not its people, but that made it no less stressful. As the hours ticked by, Amy and Bill made the best of things, talking with the other volunteers and eating what food was offered to them.

Reporters called to ask, "What do you think of your daughter being captured?"

In the evening, Amy and the others were allowed to watch the Ugandan national news on TV. That's when they learned that the Ugandan government had accused them of being "mercenaries [paid foreign soldiers] bound to destabilize Rwanda." Apparently, the timing of their flight—on the day of the coup in Rwanda—had led the Ugandan dictator to proclaim they were involved.

News stations back home buzzed with news of the detained Peace Corps volunteers. Amy's parents heard about it, too. Reporters called to ask, "What do you think of your daughter being captured?"

What are we supposed to think? Marion wondered. "We just want her to be safe," she told the reporters.

Three days after their capture, the volunteers were released. Zaire's president, Mobutu Sese Seko [moh-BOO-too SES-ay SAY-ko], had made Idi Amin see the truth: His prisoners were only volunteer schoolteachers.

Amy felt relieved. At last she and Bill were on their way to Bukavu, a city on the southwest shore of Zaire's Lake Kivu.

~ Strange New World

During those first hectic days in Africa, one piece of advice left a lasting impression on Amy. "You will see many strange and different things over the next two years," a Jesuit priest said during an afternoon of volunteer training. "Always keep a question mark in front of your eyes. Ask 'why' before you judge something you see as wrong just because it is different."

Bill and Amy pause for a picture in an African game park. "We wanted to live in a different culture," Amy recalls of their decision to join the Peace Corps.

Armed with those wise words and her basic command of French, Amy set off to teach chemistry, biology, and physics in a local high school. The students greeted her by shouting "Get out! Go away!" in three languages—French, English, and Swahili. Bill ran into similar resistance in the English and geography classes he taught.

The reason the kids didn't want their help, Amy and Bill later discovered, was that the principal and many teachers had been taking cash from the students' parents in exchange for passing grades. So students from the wealthier families weren't really interested in learning. Amy spent several distraught evenings asking, Why are we doing this?

This teaching experience was not what Amy and Bill had bargained for. But they had made a two-year commitment to the Peace Corps, and they were determined. So they continued to give lectures and quizzes. Many students refused to pay attention in class or take tests. These kids knew they could pay the principal to get passing grades.

After a year of frustration, Amy was elated when the Peace Corps education director handed her and Bill his file box full of available teaching appointments. "Find a school that needs a science teacher and an English teacher," he said, "and we'll move you."

Students at a school in south-central Zaire await the start of a class taught by Amy.

Amy and Bill chose a small rural boarding school in the south-central part of the country. In this poorer part of Zaire, some students were able to attend school only because their entire village had chipped in the money to send them. Eager to learn, the kids embraced their new teachers.

Finally, Amy thought, *I can really make a difference.* Still, she struggled with a nagging question: *What will become of the girls in my classes in a society that offers women so little opportunity?*

Though only in their teens, many of the girls had been promised by their parents as brides. Amy knew that a tough life awaited them. She thought back to the sight of a couple at a Zairian train station: The woman carried a heavy load on her head, a baby on her back, and another child in her arms. Her husband held nothing but a slim briefcase. Women in Zaire were expected to do hard labor—collect firewood, haul water, plow fields—even while caring for babies. Men led much more leisurely lives.

An African woman takes a break from working in the fields. Amy was upset by the idea that even her brightest female students might be destined for lives of hard labor.

Amy resolved to assure the girls of their worth as human beings. She wanted them to understand they could contribute to society. She reminded them, "You're special—don't ever forget that."

One day a few girls asked Amy, "If you're special, if you're an individual, why did you give up your identity by taking your husband's name? We don't do that."

Amy thought hard about what they said. In the United States, it was typical for a woman to take her husband's last name when they married. Beyond that custom, though, Amy couldn't come up with a good explanation. She decided to change her name back to Vedder when she returned to the States.

~ *Kingly Casimir*

When they weren't teaching, Amy and Bill explored Africa by truck, boat, train, and bus. With a borrowed tent, they camped in 10 national parks and discovered the continent's wildlife, from zebras, giraffes, and buffaloes to hippos, lions, and antelopes. They also experienced African habitats ranging from hot, dry deserts and savanna grasslands to cool, wet mountains. As Amy and Bill witnessed the beauty and variety of Africa's natural world, they felt a growing desire to help conserve the continent's special places.

That feeling intensified after a visit to Zaire's Kahuzi-Biega [kah-HOO-zee Bee-AY-gah] National Park. There, beneath a rain-soaked canopy of tropical trees, Amy saw wild gorillas for the first time.

Amy poses in front of the tent she and Bill slept in while touring Africa. Hitchhiking or riding buses, the couple camped in a total of 10 national parks.

"Go ahead," whispered Adrien deSchryver, the chief warden of the park. "Go sit next to him."

At first Amy thought deSchryver must be joking. But Casimir, an enormous male eastern lowland gorilla, was too intriguing for the fearless 22-year-old to resist. With a big grin, Amy began to crawl toward the kingly giant.

But the warden had been joking. He never expected a young woman to have the courage to approach such a huge creature. Horrified, deSchryver reached out, grabbed Amy's belt and yanked her back to his side.

For the next half hour, they sat 20 yards from Casimir and his family, straining to see the animals moving in and out of a thick curtain of bamboo. Every so often, a gorilla call penetrated the forest. Amy was fascinated by the apes' moves and sounds, and by their dazzling, reflective eyes staring curiously back at her.

Amy's encounter with Casimir got her thinking: *What do I want to do after the Peace Corps?* Her answer: Return to Africa to study wild gorillas.

~ What Next?

In 1975 their two-year Peace Corps stint was up. Amy and Bill flew back to the United States and began looking for a graduate school where they could launch their careers in wildlife conservation. After visiting colleges from coast to coast, they landed at the University of Wisconsin in Madison, the state capital. They settled in and started their studies, intent on returning to Africa after finishing a semester of school.

A fellow graduate student put them in touch with an up-and-coming British primatologist, Richard Wrangham. Richard had recently visited the Karisoke [kah-ree-SOH-kay] Research Center in Rwanda's Volcanoes National Park. The director of Karisoke, Dian Fossey, was studying endangered mountain gorillas. Richard believed that Amy and Bill's studies would be a valuable addition to the research being done at Karisoke and that Dian would welcome them. Amy and Bill weren't so sure; they had heard of Dian's reputation for being difficult to work with.

Who Was Dian Fossey?

In the early 1970s, Dian Fossey became world famous for her encounters with a group of Rwandan mountain gorillas that she called Group 4.

Dian had no formal training as a biologist or conservationist. She held a degree in occupational therapy, an area of medicine that helps people recover from illnesses by doing special exercises. But Dian's interest in Africa and gorillas, combined with an encounter with the famous archaeologist Louis Leakey, set her on a course to become a well-known gorilla researcher.

After establishing Rwanda's Karisoke Research Center in 1967, Dian Fossey focused her studies on gorilla behavior. She spent many hours during the gorillas' daily mid-afternoon rest watching their family interactions. One of her biggest breakthroughs came in 1970, when an adult male gorilla known as Peanuts reached out and touched Dian's hand. It was the first "friendly" gorilla-to-human contact ever recorded.

When Amy and Bill arrived at Karisoke in 1978, Dian was no longer doing much field-work. Since the slaughter of Digit, she focused instead on launching a public campaign against gorilla poaching. Two years later, Dian left Rwanda to take a position at Cornell University in Ithaca, New York. There she finished writing *Gorillas in the Mist,* a book about her experiences studying mountain gorillas.

Magazine articles and documentary films made Dian Fossey and the Virunga mountain gorillas famous worldwide. Had it not been for the media attention, the Rwandan gorilla population would not have become known to the world.

When Richard presented Amy and Bill with a written statement from Dian saying that she was interested in learning more about their work, the couple revised their research proposal to fit the situation in Rwanda. Then in October 1977, they met with Dian at a hotel restaurant in Chicago. After dinner, Dian read their research plans. "Working with the local people is hopeless," she commented, but she accepted their proposal and set a tentative date for Amy and Bill's arrival in Rwanda in a few months.

Digit *(above)* was made famous by TV documentaries and magazine articles featuring Dian Fossey's work with mountain gorillas. In February 1978, Digit was killed by poachers.

On February 3, 1978, the CBS Evening News reported that poachers, illegal hunters, in Rwanda had killed a mountain gorilla named Digit, made famous through Dian Fossey's research. By 1978, Amy and Bill had several semesters of graduate school under their belts. They also had secured enough research funding from the Wildlife Conservation Society, based at New York City's Bronx Zoo, to afford a diet of rice and beans in Rwanda for 18 months.

The two headed straight for the Karisoke Research Center in Volcanoes National Park—the forest where poachers had brutally murdered Digit.

It felt good
to be immersed in an
African culture
once again

and to be surrounded by this
stunning
landscape.

3

In the Land of a Thousand Hills

The wind blew Amy Vedder's short, chestnut-colored hair as she and Bill bounced across the Rwandan landscape in the back of a rickety pickup truck. Just a few weeks had passed since Digit's murder. Amy had learned Swahili [swah-HEE-lee] in the Peace Corps, so she was able to make sense of the conversations swirling around her. The truck was filled with more than a dozen Rwandans, but none of them were talking about Digit. Amy understood why. Most people in Rwanda had never heard of the famous mountain gorilla in their midst.

Summits of the Virunga Volcanoes *(opposite)* mark the international borders of Volcanoes National Park in Rwanda, where Amy studied mountain gorillas on the slopes of Visoke *(above).*

Outside Africa, however, Digit was almost a household name. Worldwide, thousands of people had read magazine articles and watched TV documentaries about the gorilla and his family. When the news broke that the young silverback had been struck down in his prime by poachers toting spears and machetes, postcards poured into Rwanda from all over the world. The messages urged the government to protect the remaining mountain gorillas.

A lot of people will be paying attention to what happens to these gorillas, Amy thought as the truck bumped along. The idea intimidated her a little. Nonetheless, it felt good to be immersed in an African culture once again and to be surrounded by this stunning landscape.

On the flight into Rwanda's capital, Kigali, Amy had peered down at Rwanda's jagged surface, seeing for the first time why Rwanda is called "The Land of a Thousand Hills." Now, from the truck bed, she had a 360-degree ground-level view of the countryside. Leafy coffee and banana plantations dotted the hillsides, connected by fields of beans and sweet potatoes. Four hours into the drive, the country's biggest hills—the Virunga Volcanoes—appeared one by one on the horizon. Ranging in height from 11,500 feet to nearly 15,000 feet, the mountains have symbolic names: Karisimbi [kah-ree-SEEM-bee], the "cowrie shell";

Gorilla Species

The world's first gorillas emerged between 6.2 million and 8.4 million years ago in the damp, low-lying forests of western Central Africa. Migrating eastward, these early apes encountered the mighty Congo River and traveled its northern bank for hundreds of miles and countless generations. Eventually, a few gorilla families split off to settle at the top of the Western Albertine Rift Mountains.

Throughout this time, the Earth's climate was changing. Regional trees and plants that could not adapt to the new climate were replaced by "fitter" species. This environmental change also affected animals. Over tens of thousands of years, the gorillas evolved into separate species, *Gorilla beringei* and *Gorilla gorilla*. Under these main groups, scientists recognize four subspecies, or types, of gorillas: the eastern lowland gorilla, the western lowland gorilla, the mountain gorilla, and the Cross River gorilla. Each subspecies possesses physical characteristics suited to its habitat.

The first gorillas that Amy saw in the wild were eastern lowland gorillas *(Gorilla beringei graueri),* which inhabit the massive forests of the Democratic Republic of Congo (DRC), formerly Zaire. Eastern lowland gorillas *(below)* have longer faces and broader chests than western lowland gorillas *(Gorilla gorilla gorilla),* which live in the low-lying areas of Gabon, Cameroon, and the Congo Republic. Eastern lowland gorillas are black all over, while the top of the western lowland gorilla's head is often brown.

Visoke [vee-SO-kay], the "watering hole," named for its crater lake; the jagged Sabyinyo [sah-bi-EEN-yo], "teeth of the old man"; Gahinga [gah-HEEN-gah], the flat-topped "hoe"; and Muhavura [moo-hah-VOO-rah], known as the "guide" because its height and location make it such a widely visible landmark.

These five volcanoes form the international borders of Volcanoes National Park, created in 1925 as a protected reserve for mountain gorillas. Amy couldn't wait to explore the park's valleys and steep slopes for answers to some puzzling gorilla mysteries.

The mountain gorilla *(Gorilla beringei beringei)* is the subspecies that Amy studied in Rwanda *(right)*. They have longer hair than other gorillas, giving them a shaggy, almost cuddly appearance. Sadly, only two populations of mountain gorillas, each with about 300 to 400 animals, survive in the forests of Uganda and Rwanda and just over the border in DRC.

Nearly 300 individuals of the Cross River gorilla *(Gorilla gorilla diehli)* live in a few isolated areas along the border between Nigeria and Cameroon. Although this subspecies does not appear to differ from other western gorillas in body size, the animals have smaller jaws and shorter skulls.

Today, both eastern and western lowland gorillas are considered endangered, or threatened with extinction, a fact that concerns many people in Africa and around the world.

No one knew exactly how many mountain gorillas were left in Rwanda, but it was certain that the species was critically endangered. In 1973, the last time anyone had counted the animals, the Rwandan population stood at only 250 to 275 gorillas. (A second population in Uganda contained a few hundred more.) Now, with Digit gone, Amy was acutely aware that the Rwandan population had just fallen one number lower.

A few researchers were scrambling to learn everything they could about mountain gorillas before they went the way of the dinosaurs and became extinct. It was only a matter of time, these researchers were convinced, before the animals disappeared forever. Unwilling to stand by and document the gorillas' demise, Amy and Bill wanted to find a way to save them.

There was still so much to learn about the fascinating creatures. Less than 100 years had passed since German explorer Oscar von Beringe reported seeing some tall, "manlike" creatures in the steep

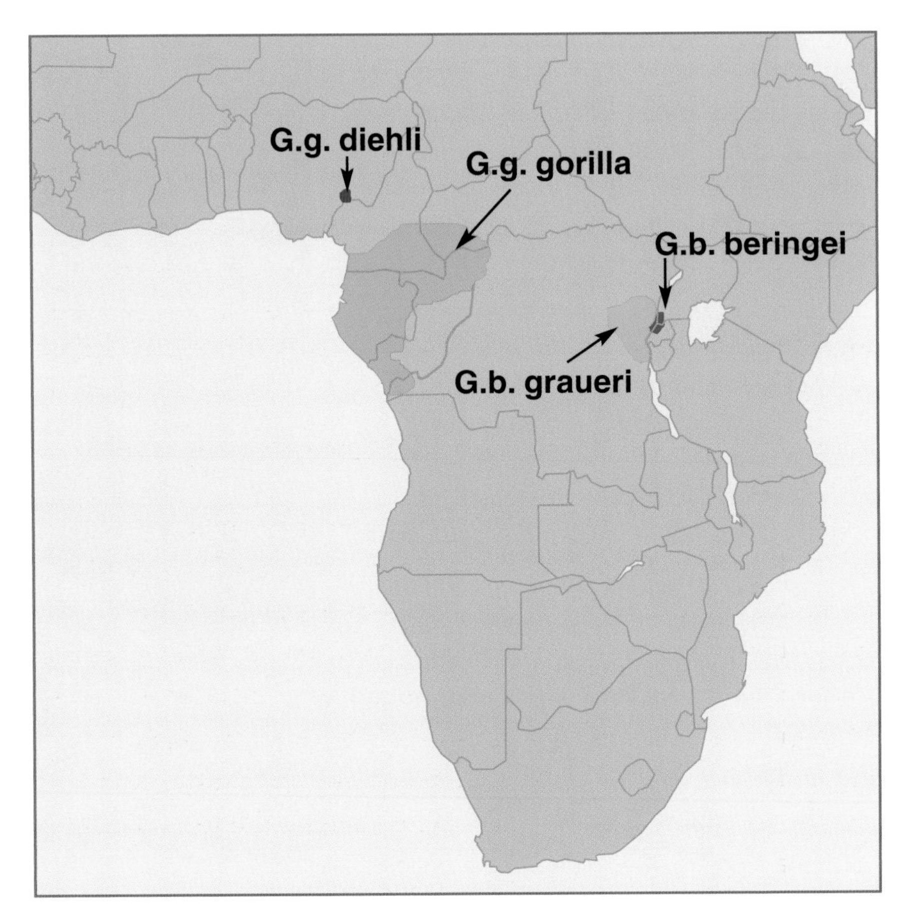

Wild gorilla populations can only be found in isolated regions of central Africa. All four of the subspecies that live there today are endangered, meaning they are at risk of becoming extinct.

Central African forests of Rwanda in 1902. Less than 20 years had passed since wildlife biologist George Schaller debunked the myth that gorillas were man-eating monsters like King Kong. And only 10 years had passed since Dian Fossey showed how similar gorillas are to humans in behavior.

Like Dian, Amy was interested in gorilla behavior. But merely learning more about it, she knew, would never save the dwindling species. Instead, she planned to focus on the mountain gorillas' ecology, the relationship between the animals and their environment.

As a match to Amy's studies, Bill aimed to find out how many gorillas still survived and to understand the pressures that were driving Rwandans and their leaders to destroy the gorillas' forest home.

~ *At Home in the Hills*

When Amy and Bill reached the small town of Ruhengeri [roo-hen-GEH-ree], near the base of Visoke, it was late afternoon. They arranged a ride with another truck driver, who agreed to take them to the trail leading to Dian Fossey's camp. From there they would have to journey on foot—about an hour's climb.

At the trail head, roughly 9,000 feet above sea level, Amy got out of the truck, finding it slightly hard to breathe. The air would be even thinner at 10,000 feet, where Dian's field station sat in the mountain saddle between the volcanoes Karisimbi and Visoke (hence the station's name: Karisoke).

Amy and Bill slung their backpacks over their shoulders and started hiking up the narrow trail that cut through a rock wall of lava, rubbed smooth by passing elephants. As the trail climbed the southern side of the volcano, it pierced an emerald-green oasis kept wet by the region's abundant rainfall, then meandered through grassy clearings and crossed muddy streams. Amy huffed and puffed in the thin mountain air. Finally, in the distance she saw the six cabins at Karisoke scattered among the forest's giant Hagenia trees.

Amy and Bill lived in this small tin house while working at the Karisoke Research Center. The house had no running water; a wood stove provided heat. Water for bathing had to be heated over an open fire *(below)*.

On the camp's edge sat the cabin that Amy and Bill would call home for the next 18 months. With its corrugated-tin walls and roof, the shack promised little warmth from the region's near-freezing nighttime temperatures. Inside, though, it had all the necessities. There was a wood stove, a desk where they could write field notes, and a small table for a single kerosene burner. Amy and Bill would need the burner to heat whatever one-pot meals they could invent. Several long boards formed another table for dining. Some shelves for their books and food staples, a second room with a bed and a dilapidated dresser, and an outhouse in the back completed Amy and Bill's new home.

The cabin grew chilly after sunset, and Amy piled sleeping bags on the bed to stay warm. The next morning she and Bill woke early and brewed a quick cup of tea. Then they raced off to meet the young British biologist who would take them to the area where Digit's family had been spotted the day before.

Following the trail that gorillas leave—trampled vegetation, neat mounds of animal dung, half-eaten plants—they traced the group's wanderings. After a few hours of tracking the signs as they walked, the researchers almost bumped right into Digit's group, silently resting and feeding in a small clearing. Amy watched the gorillas in amazement as they not only tolerated her approach but observed her with curiosity. Could they be just as intrigued by her as she was by them?

Amy knelt on the ground, sitting still so she wouldn't disturb the apes. Moments later, a two-year-old male gorilla named Kweli [KWEH-lee] approached. Kweli shyly looked away and touched the hem of Amy's jeans. Then he twirled back to his mother's side.

Amy marveled at Kweli's deep brown eyes and the peacefulness of the group. *Why don't you run?* she wondered. *Don't you remember the poachers? Or the sound of Digit's screams mixed with the thud of their machetes?*

By first light,

she was on the gorillas'

trail.

4

Midnight Voyage among the Volcanoes

Amy had enjoyed her visit with Kweli and his mother. But in the weeks after her arrival at Karisoke, she started investigating the lives of another gorilla family, Group 5—one that researchers hadn't yet studied much. Each day she woke before sunrise and put on her standard uniform: a turtleneck, heavy flannel shirt, and wool sweater paired with loose-fitting cotton pants, wool socks, and Army surplus jungle boots. Before she left the cabin, Amy grabbed a bite of bread with peanut butter and guzzled a cup of warm tea. By first light, she was on the gorillas' trail.

To follow the gorillas in Group 5 all day, Amy had to gain the acceptance of Beethoven *(above)*, the reigning male silverback. Now and then Beethoven made it clear that Amy was not an invited guest.

Rwelekana [Rway-lay-KAH-nah], a Rwandan gorilla tracker, often helped Amy locate Group 5. Moving quietly through the jungle, he and Amy talked softly about the gorilla signs they found. From time to time, Rwelekana paused to pull a wire snare from the wet underbrush of a path. Similar traps had seriously injured several gorillas in the forests surrounding Karisoke. The hunters who hid the snares meant no harm to the apes; they were only trying to catch forest antelopes. Rarely are gorillas hunted for food in Rwanda, a country where it is taboo, or socially unacceptable, to eat gorilla meat or any other type of primate meat.

Hunters hoping to catch antelopes placed this illegal wire snare on an animal path in Volcanoes National Park. Occasionally such traps snag a mountain gorilla instead.

Amy knew the hunters were foraging for food to feed their families. Still, they weren't supposed to hunt inside the national park. To protect the gorillas and other animals, Rwelekana and Amy collected the snares they found and stuffed them in their backpacks.

~ Lost Ape

Despite their efforts, it wasn't long before Amy saw a mountain gorilla that had fallen victim to a hunter's snare. In early March 1978, an official with Rwanda's national park service told Bill that park guards in neighboring Zaire had seized a young female gorilla from poachers trying to smuggle it out of Rwanda. The poachers had bought the animal from an antelope hunter who had trapped it by mistake. Now they planned to sell it at a profit, probably to an exotic pet collector.

The Congolese park staff didn't know what to do with an injured gorilla. They were holding her in a dark, smelly shed at park headquarters when Amy and Bill arrived with Dian Fossey several weeks later. The guards opened the shed and poked long poles at the frightened gorilla, forcing her to limp into the sunlight.

Amy felt sick at the sight of the poor little ape, its hair streaked with its own diarrhea. It was decided it would be best if she stayed with the animal while Dian and Bill arranged to transport it back to Karisoke.

"She must be moved to a cleaner room," Amy told the park guards. They seemed puzzled by the idea. Why would a gorilla need a cleaner place to sleep? But they agreed to shift the animal to another shed.

Coaxing the gorilla into her new room, Amy closed the door and knelt on the floor. She thought of the abuse the animal had suffered at human hands and tried to put it at ease by making a noise deep in her throat: *"Hhhhggmmm, mghaaaah."* Gorillas make just such rumbling sounds when they come in contact with one another. Hearing the familiar noise, the gorilla climbed into Amy's lap.

Amy was shocked. Tears streaked down her cheeks as she hugged the frightened creature close.

Amy holds Mweza, a young female gorilla she and Bill nursed back to health after it was confiscated from poachers in March 1978.

~ *Taking Mweza Home*

Holding the gorilla in her lap, Amy could feel how skinny she was. Judging by the length of the gorilla's body, Amy estimated she was about four years old, yet she weighed only as much as a two-year-old. The terrified gorilla obviously had not been eating much; the park guards had fed her powdered milk mixed with dirty water, which made her sick with diarrhea. Part of a metal snare protruded from the gorilla's swollen and infected left ankle.

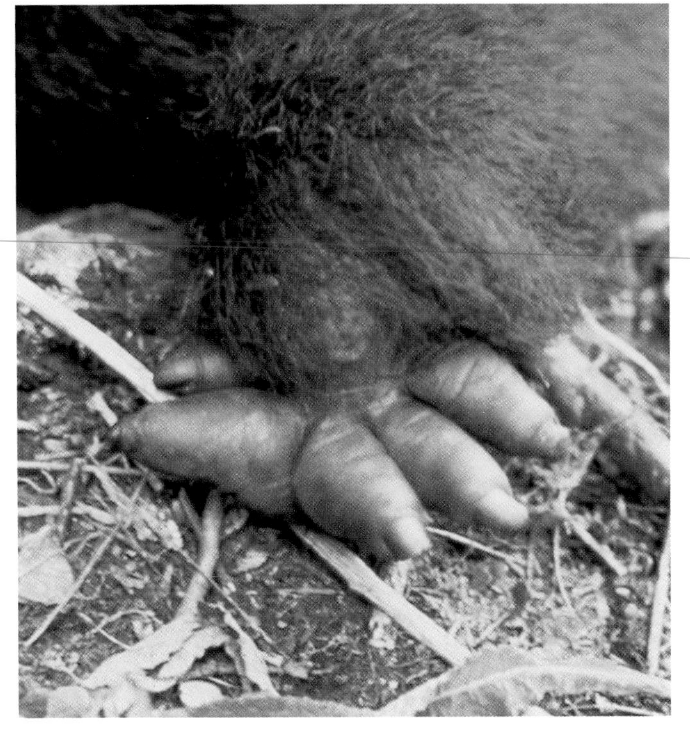

The foot of this gorilla was permanently injured by a metal snare. Such injuries often leave the animals handicapped for life.

Amy didn't know if the young ape would survive, but she planned to improve its chances. The mountain gorilla population could not afford to lose another member, especially a female who might one day have babies of her own. Thinking of the children's book character *The Little Engine that Could*, which chugged up a steep hill by saying "I think I can, I think I can," Amy named the gorilla Mweza [MWAY-zuh]. The word means "can do" in Swahili. "I want her to fight against all odds and survive," Amy said.

Amy stayed with Mweza day and night while Dian and Bill tried to persuade the authorities to let them move the injured gorilla to Rwanda in their custody. In the afternoons she walked Mweza around the park grounds and reintroduced her to the plants she had eaten in the wild. At night Amy kept Mweza and herself warm by sleeping curled up with her on a straw mattress.

After two full days, Zaire's President Mobutu agreed that Mweza could be taken to Rwanda—but only if Dian promised to release the gorilla into the wild if she recovered. If not, Mweza's body was to be returned to Congo. Dian gave her word.

Crossing the border into Rwanda with an undocumented gorilla would be tricky, however. The best way to take Mweza home, Dian decided, would be for Amy and Bill to carry her over the volcanoes while Dian drove back to the park. Toting Mweza like an African child in a cloth sling on their backs, the couple set off on the 16-hour climb.

Although it towers 14,557 feet high, cloud-shrouded Mount Mikeno could not stop Amy and Bill from carrying Mweza to safe ground.

To help Amy and Bill find their way back to Karisoke, the park officials sent eight armed guards with them. Marching through the midafternoon mist, the couple took turns lugging Mweza over Mount Mikeno. Giant stands of bamboo and deep gorges distracted them from the pain of Mweza's sack cutting into their shoulders. Amy and Bill would have liked to have camped beneath the forest's enormous Hagenia trees. But Mweza was unlikely to survive the evening chill, so they pressed on.

Angry that they couldn't stop for the night, the guards grumbled as they led Amy and Bill toward Dian's camp. Darkness crept over the forest, and the guides—perhaps in protest—claimed to be lost. The exhausted group stumbled on through wet, heavy underbrush. Finally, just before midnight, Amy and Bill spotted the lights of Dian's cabin in the distance.

~ A Cabin for Three

Over the next few days, Amy, Bill, and Mweza settled into a new routine. In normal circumstances, the couple would have hesitated to allow a wild gorilla to bond with them. But Mweza was sick, so they took turns sleeping with her on the floor. By staying close, they could keep an eye on Mweza's condition. They also hoped their presence would ease her anxiety.

In the afternoons, Amy continued to reintroduce the frail gorilla to natural foods such as wild celery and blackberries, which quickly became Mweza's favorites. She also fed Mweza a liquid antibiotic to fight the painful infection in her wounded ankle.

Mweza was in caring hands. Amy had some practice in tending to sick animals. At home in upstate New York, she had once saved an ailing puppy that someone had brought to her father for help. No one in Amy's family thought the puppy would live, but she had nursed it back to health by feeding it from an eyedropper. Now she hoped to work the same magic on Mweza.

Amy pushed down on Mweza's chest. Bill breathed into her mouth. Eventually, Mweza started breathing on her own.

Sure enough, Mweza began to perk up. She surprised Amy—on her birthday—by eating some blackberries all on her own. Indoors, Mweza became rambunctious. One day she climbed up on a desk and threw papers and books onto the floor. She made a mess, but Amy just laughed. It was a sure sign that Mweza was growing stronger.

~ *Call 9-1-1*

Mweza still needed medical care for her injured ankle. A French doctor at a nearby hospital had offered to help Dian and the Karisoke staff with sick gorillas, but Dian refused to let him tend to Mweza. She and another of the hospital's doctors had carried on a brief love affair that ended badly and now Dian's pride kept her from seeking help from the hospital staff. She refused to call on them, vowing, "I'll never take *any* gorilla there."

With the wire snare still stuck in her ankle, Mweza started to weaken. Then her condition took a sudden turn for the worse. Late one night she stopped breathing.

Kneeling on the floor of their cabin, Amy and Bill frantically tried to revive the young ape using CPR (cardiopulmonary resuscitation). Amy pushed down on Mweza's chest. Bill breathed into her mouth. Eventually, Mweza started breathing on her own.

But not for long. Amy and Bill sprang back into action with more primate CPR. When Mweza was revived a second time, she must have been startled to see two humans hovering over her because she bit Bill's lower lip. Bill was bleeding, but his lip was intact. The next time Mweza stopped breathing, Bill wisely blew air into her nose rather than her mouth. That method worked even better.

Amy and Bill resuscitated Mweza several more times during the night. Then, just before dawn, Mweza failed to wake up. In the pale morning light of Amy and Bill's cabin, the gorilla population clicked down one more number.

Amy and Bill

were more determined

than ever

to stop the mountain gorillas

from becoming

extinct.

5

Amy the Ape

Amy and Bill were angry. Angry that Karisoke lacked a medical system to treat gorillas injured by humans. Angry at Dian Fossey for not calling the doctor who had offered to help gorillas like Mweza. Angry at the people who had plucked Mweza from the forest.

After Mweza died, Amy Vedder buried her sorrows in her fieldwork.

But they were also more determined than ever to stop the mountain gorillas from becoming extinct. They faced two big challenges: protecting gorillas from hunters and poachers, and making sure the population had enough forest food and land, or habitat, to survive.

Amy focused her efforts on the second challenge. She needed to write down everything Group 5 ate and where they ate it. With that information she could figure out what type of habitat—and how much—the gorillas needed.

At the same time, Bill began figuring out how many gorillas were left. He trekked up and down the volcanoes, counting every gorilla he could find. By the fall of 1978, Bill knew there were still about 260 gorillas. The population was stable—and its high number of young animals was a good sign for the gorillas' future.

~ *Sleepless Nights*

In the weeks after Mweza's death, Amy often woke up before dawn feeling anxious. With Bill away doing his gorilla census work, she was alone in the chilly cabin. But that didn't bother her as much as the fear that she might not find Group 5 in the morning.

Following Mweza's death, Dian Fossey had abruptly turned on Amy and Bill, blaming them for the ape's demise. In addition to cutting off their supply of firewood and stopping their daily delivery of a 20-liter can of warm water for bathing, Dian told Rwelekana and the other Karisoke trackers not to help Amy find Group 5. Life at 10,000 feet became less comfortable, but Amy and Bill adapted by cutting their own wood and heating their bath water in a tea kettle.

Despite the harsher conditions, Amy resolved to stick with her research and track the gorillas on her own. It was essential that she locate the gorilla family each day. If not, she would be unable to collect enough data to draw any meaningful conclusions about the gorillas' habitat needs.

Fortunately, Amy's weeks with Rwelekana had taught her a lot about gorilla tracking. The dense vegetation covering the forest floor didn't show footprints, but the gorillas left small clues for

Amy transcribes her field notes by lantern light after a long day of tracking gorillas in the cold rain.

Amy to follow. A leaf with dew on its underside, for instance, meant the gorillas had overturned it as they passed by that morning. That single telling detail could mean the difference between Amy going straight to the gorillas' location or walking in circles for hours.

Gorilla dung held all sorts of clues as well. Dung that was dry or had insects crawling beneath it indicated the trail was old. Dung that was warm when Amy felt it with the back of her hand meant the gorillas were nearby. (Amy used the back of her hand just in case she absentmindedly touched her fingers to her mouth later.)

Gorilla dung is a reliable way to track the animals that left it behind. Dry dung means the animals are long gone; warm dung means the trail is still hot.

As she roamed the forest looking for signs of the gorillas' route, Amy couldn't help recalling the words of other researchers: "You will never be able to record everything the gorillas eat. They will never let you stay with them all day."

~ *Trying to Fit In*

Although it was unwise to stare at length into the eyes of the gorillas she studied, Amy sometimes found them peering curiously at her.

Amy knew she would have to break new ground. No one had ever followed a gorilla family from sunrise to sunset. All prior studies had taken place while the animals paused for their midday rest.

Once she got the hang of finding the gorillas, Amy discovered the others were right. The shaggy apes really didn't like having a human hanging around all the time. Often, they let her know she wasn't welcome with vocal warnings that sounded like deep coughs. Twice, Beethoven, the 400-pound silverback of Group 5, grabbed Amy's arm and squeezed it, as if to say, "It's time for you to go now."

Amy took Beethoven's warnings seriously. She tried to make her presence less intrusive by acting—as best she could—like one of the gang. She walked like the gorillas: hunched over, with her arms close in. She communicated as they did, too, giving belch vocalizations as she slowly approached the group. Then she would squat nearby and try to blend in by mimicking the gorillas' actions. If the gorillas were feeding, Amy collected handfuls of plants, held them to her chin, and pretended to chew.

As she performed this masquerade, Amy took notes on small pieces of paper. Before leaving for Africa, she had bought waterproof notebooks that would protect her field notes during Rwanda's rainy season, when daily downpours are the norm. But someone had stolen the trunk containing those notebooks at the airport. So Amy had to make do with sheets of paper attached to a clipboard and covered with a plastic bag. Many days were so cold and wet she had to force her fingers to work the pencil inside the bag.

Amy was careful not to stare at the gorillas as she studied their movements. If she looked too long, they might take it as a threat and charge. Gradually she learned to identify each of the 14 members in Group 5 by the animal's body type and posture. Within six weeks, the gorillas seemed to be growing familiar with Amy, too. They tolerated her presence more and more.

As Amy tagged along with Group 5, she learned that they stuck to a home range of about five square miles. Dense bamboo stands, rare concentrations of blackberries, flowing streams, and elephants were part of the richly varied landscape within this home range.

Amy carefully records the activities of three mountain gorillas. Each day she would focus on one animal, noting how it spent its time, what it ate, and which gorillas it played with.

Amy developed simple methods for recording the foods eaten by the gorillas. Each day, for roughly five hours, she watched a single member of the group, noting everything this "focal" gorilla ate. She scribbled notes about food quantities, the amount of time the gorilla spent eating each item, and what other foods were within reach but

not chosen. Some days as her focal animal moved about eating different types of vegetation, Amy selected equal amounts of the preferred plants and stuffed them in a garbage bag. At the end of the day, she took the bag back to her cabin and weighed its contents. Then she carefully dried the plants above her woodstove so she could evaluate what nutrients the food provided.

Amy sometimes collected equal amounts of the foods—celery *(above)*, blackberries, bamboo—that a single gorilla ate in the course of a full day.

After one afternoon of duplicating Beethoven's meals, however, Amy went home empty-handed. As she respectfully looked on, the silverback had walked up and swiped her bag! Amy didn't dare object or snatch it back from the kingly giant.

Amy just watched dispiritedly as he sat down a few feet away and stuffed wads of plant data into his mouth. She felt comforted that Beethoven seemed to approve of her choices—he ate every bite. Then he dropped the bag and wandered off for a nap.

~ Gorilla Games

It didn't take long for Amy to learn the individual personalities of all the gorillas in Group 5. There were Beethoven and Icarus, the two silverbacks; Effie, the confident matriarch and her offspring, Puck, Tuck, and Poppy; wrinkly old Marchessa, who had a young son, Shinda, an older daughter, Pantsy, and a bold granddaughter, Muraha; little Pablo, the clown, and his mother, Liza, who would later abandon him; and two adolescents named Quince and Ziz.

Amy especially enjoyed watching the youngest members of Group 5. They were curious and full of energy. While the older gorillas settled in for their late-morning naps, the youngsters chased each other in circles in a game of gorilla tag or raced one another up a fallen tree as if they were playing King of the Mountain. Often the youngsters wrestled one another and let out chuckles as they sparred.

The youngsters in Group 5 tried to bait Amy into playing with them. Some mornings started with Pablo dropping from the trees onto Amy's back. At about 65 pounds, the four-year-old gorilla could not be ignored. If Amy tried, the trickster playfully grabbed her backpack and pulled her to the ground.

As best she could, Amy resisted the temptation to play with Pablo and the other young gorillas; she didn't want her interactions to change the animals' natural behaviors. But sometimes she got stuck in the middle of their games. One day two young females, Muraha and Poppy, began chasing each other in circles around Amy, pulling vines as they ran. When the two were done with their game of ring-around-the-researcher, Amy was wrapped up in vines like a mummy. She had to use her pocketknife to cut herself free.

The things Amy wore and carried captivated the young gorillas. They liked to untie her bootlaces or steal her binoculars and

Noseprints

Whenever gorilla researchers see an animal they've never met before, they quickly sketch the animal's "noseprint"—the pattern of deep lines that crease the skin above a gorilla's nose. Like a human fingerprint or a jaguar's spots or a zebra's stripes, each gorilla's noseprint is unique.

The Karisoke field station kept a collection of noseprint sketches on file. Researchers could use these to identify and monitor animals that did not belong to long-term study groups.

backpack. (Fortunately, none of them ever figured out how to work the bag's zipper.) Yet Amy was always surprised by how gingerly the boisterous gorillas handled her things. They passed her binoculars around as if they were fine china. Six-year-old Tuck was especially fascinated by the freckles on Amy's arms. She tried to pick them off with her fingernails and vacuum them up with her lips.

~ Daily Life at the Top of the World

Amy grew accustomed to her adopted family. Within a few months, the gorillas were completely accustomed to her, too. Amy spent nearly every day with Group 5. On rare breaks from her research, she caught up on her field notes or rushed down to the capital, Kigali, to renew her three-month tourist visa (a document that gave her permission to stay in Rwanda).

Gorilla Science

All scientific studies start with a hypothesis, an idea that can be tested by observation or experiment. After reading journal articles and kicking ideas around with her graduate school professors, Amy decided on a few hypotheses to test while researching gorillas in Rwanda. One of them was the notion that gorillas spend more time feeding in areas with a wide variety of foods. Having many foods to choose from, Amy reasoned,

allows the apes to get the nutrients (proteins, carbohydrates, vitamins, and minerals) they need to stay healthy. Beyond the scientific value of Amy's hypothesis lay a conservation angle. If she could figure out which food-rich areas the gorillas visited most often, she could argue that those areas should be off-limits to other uses, such as cattle grazing.

To test her hypothesis, Amy tracked Group 5 closely during the 18 months she lived at the Karisoke field station *(above)*. Her only tools were a compass, an altimeter (a device that measures elevation), and a topographic map, which shows in detail the features of a landscape. Using these instruments, Amy recorded Group 5's wanderings from dawn to dusk. As Amy tailed the gorillas, she kept track of how far, and in what direction, the animals walked each day. She also noted details about the habitats through which the group passed as it moved from one feeding zone to the next.

After long days of fieldwork, Amy returned to her cabin and transcribed her field notes onto larger data sheets. Later, she sketched crude maps of Group 5's Visoke home range atop a grid *(right)*. Each time Group 5 visited an area, Amy drew a slash through that area on the grid. More than one slash—an x or *, for example—represented multiple visits from the group.

To survey the food types within each region of the grid, Amy collected plant samples from areas chosen at random. She noted both the quantity and quality of plant foods present.

When she returned to the University of Wisconsin, Amy used a special machine to transfer her field data onto punch cards. She then fed the cards into a computer—or the slow, balky machine that passed for one in 1978—which read and sorted the statistical data. After hours of processing time, the computer noisily spat out a printed report.

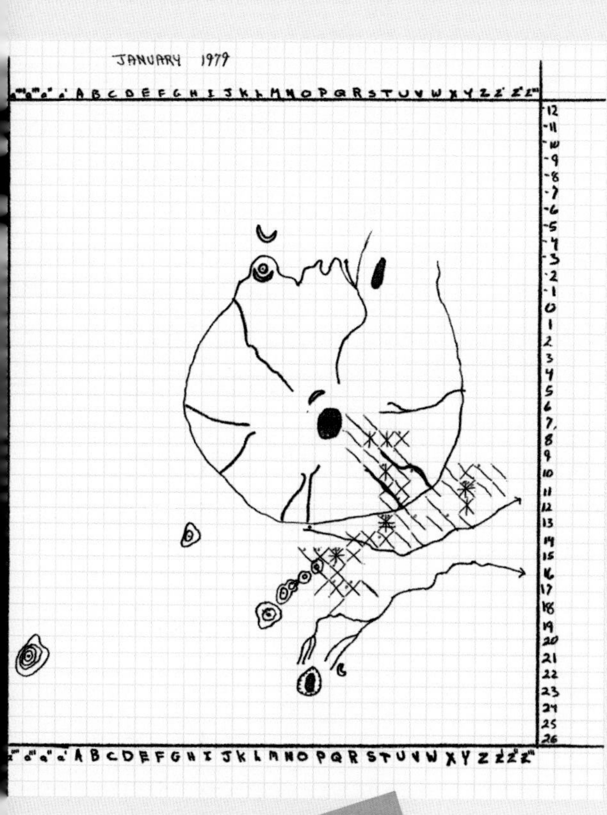

Amy's data analysis confirmed her hypothesis: The gorillas preferred areas that contained a high variety of foods. Two areas in Volcanoes National Park were crucial to the gorillas' survival. In addition, the gorillas also sought out areas that abounded in bamboo, a key source of protein for the apes.

After double-checking her data collection, analysis, and conclusions, Amy wrote up her results in a scientific paper illustrated with charts and maps. She sent the paper to a prestigious scientific journal, the *American Journal of Primatology,* and waited to see if the editors would publish it. Amy was ecstatic when she learned that the journal would feature her findings *(below)*. Better yet, her research helped identify habitats vital to the mountain gorillas' survival. Those areas were saved from conversion to cattle pastures when Amy cofounded the Mountain Gorilla Project *(page 53).*

Gorilla Ranging Patterns 81

Other gorilla groups were present to the north, west, and south and may have been
important factors in determining these boundaries of the home range.
Group 5 did not use the home range in a uniform manner; as shown in Figure
3, the frequency of visits in quadrangles ranged from 1 to 45 with a concentration of
use shown in two core areas, one smaller than the other. A similar pattern was
derived from data on total duration of quadrangle use. Cumulative use increased
logarithmically with rank order of quadrangle, such that 25% of the group's time,
as measured by durational use, was spent in 5.5% of the quadrangles.

LMNOPQRSTUVWXYZ Z'Z"

6 7 8 9 10 11 12 13 14 15 16 17 18 19 20

41–45
31–40
21–30
11–20
6–10
1–5
NOT VISITED

Fig. 3. Frequency of use of quadrangles in the home range of Group 5, showing a differential use of space
with two core areas. (Quad coordinates follow conventions of all Karisoke Research Center publications;
grid system as in Fig. 1.)

Yearly Ranging Patterns and Spatial Distribution of Food

Table II gives results of correlation analyses conducted for each of the two
ranging parameters vs. five food availability measures. Three of the correlations
were statistically significant: positive correlations between both the abundance and
frequency of highly proteinaceous foods per quadrat and time spent per visit, and a
positive correlation between diversity of foods available per quad and frequency of
use of a quadrangle. Each of these food measures represented some quality of
available foods, not merely abundance or frequency of all foods together. In fact, all
of the six correlations between food quality parameters and those of ranging were in
the direction expected (positive). However, no consistent relationship is found between availability of all commonly eaten foods and ranging measures, correlations
being either positive or negative.

Occasionally, Amy made a phone call to the United States from the city. But that required staying overnight with friends and missing two days in the field with the gorillas. To call the United States, she first had to dial the local Rwandan operator, who connected the call to an operator in Belgium. (Rwanda had been a Belgian colony.) Then, often more than 10 hours later, the Belgian operator rang Rwanda back when the call was connected to the United States.

Given the cost and complexity of the Rwandan phone system, Amy communicated with her family mostly by letters; these were picked up and dropped off by porters, who climbed the trail to Karisoke with supplies twice a week.

Amy spent more than 2,000 hours observing Group 5 to learn about their feeding habits. Often the weather was wet and cold. But the rewards were worth it.

Amy's isolated lifestyle strengthened her bond with Bill. On nights when Bill was home from his census work, they cooked dinner together while she updated him on the Group 5 youngsters' latest tricks.

Those dinners could go on for hours. Amy never ate lunch when she was in the field, except for an occasional piece of celery with the gorillas. When she got back to the cabin at dusk, she was famished—and frozen. Dinner usually began with a cup of warm tea, followed by spicy peanuts sizzled in hot oil. Then Amy would prepare one of the meals on her revolving menu: spaghetti and tomato sauce with onions, rice and beans, chili, eggs, vegetable stew, or Bill's favorite, "fake" Reuben sandwiches. The Reubens contained melted cheese and homemade sauerkraut that Amy had learned to create by alternating layers of cabbage leaves and salt in a glass jar. The parts missing were the traditional corned beef and Russian dressing.

~ *Just One of the Gang*

One night Amy barely made it home for dinner by the time the forest grew dark. It's never a good idea to be out after sunset in the Virungas. That's when Cape buffaloes—one of Africa's most dangerous animals—are active. Few people have stumbled upon a Cape buffalo and walked away unharmed.

Like the female gorillas in Group 5, Amy was ever mindful of Beethoven, giving the 400-pound silverback as much space as he needed.

Amy wasn't late on purpose. She had been hurt in a scuffle with Beethoven. Spotting Effie quietly enjoying some blackberries, Beethoven decided to help himself. He marched over to Effie, shoving past Amy and knocking her off balance. As Amy fell to the ground, she wrenched her ankle against a tree root. Pain shot up her leg.

Already Amy had stayed with the gorillas later than usual. She didn't know if she would be able to reach camp before dark. Pulling herself together, she used her machete to fashion a cane from a branch, then began hobbling back to camp. The path was stony and steep, and Amy stumbled over the rough terrain. At last she made it, just minutes before Bill was to launch a search party.

Amy didn't blame Beethoven for her injured leg. He was only doing what came naturally. After all, she was starting to feel like a member of his clan. She had even been formally "inducted."

One day during the family's midmorning siesta, the group was sprawled out on the ground in its usual daisy-chain formation, each gorilla touching another. Amy was sitting nearby when young Ziz rolled over, looked up at her, and rested his hand on her arm, joining her to the chain.

It was a special gift to the wildlife biologist who had worked so hard to fit in.

Uncle Bert's family,
known as Group 4,
had already lost
one ***silverback****—Digit.*

Now their ***leader***
was dead.

6

UNDER ATTACK

July 24, 1978

Bill:

Uncle Bert has been killed. We don't know where the other gorillas are or if they are alive. Please come back to camp as soon as you can.

—Dian

Uncle Bert *(left),* the majestic silverback, was the leader of Group 4 before his brutal murder. Protecting bamboo plants *(above)* and the rest of the gorillas' habitat became Amy and Bill's mission.

Bill went numb as he read the note that had been delivered by a porter. He had just finished an afternoon of census work and was standing on the northeast flank of Mount Visoke, more than a three-hour hike from Dian's camp.

Bill arrived at Karisoke later that night, tired and sweaty from covering the hilly terrain as quickly as possible. Amy met him at the door of Dian's cabin. They hugged in a sad, wordless embrace, then went inside to talk about what had happened.

At about 8:00 that morning, a Karisoke researcher had found a headless pile of hair and flesh on the forest floor. Uncle Bert's body was still warm, suggesting the poachers were not far off. Maybe they had scattered at the researcher's approach.

Uncle Bert's family, known as Group 4, had already lost one silverback—Digit. Now their leader was dead.

None of the other males in the group of 12 gorillas were old enough to take over—if they were even still alive. After discovering Uncle Bert's body, one of the Karisoke researchers followed Group 4's trail up Visoke's slopes. He found a number of the gorillas resting and was surprised by how peaceful they looked in the wake of such a brutal attack. By nightfall, however, he couldn't account for all of the remaining group members.

Amy and the other Karisoke researchers learned about Uncle Bert's murder only after returning from a full day in the field. As the clock ticked toward midnight, she felt helpless and exhausted. Deciding there was nothing they could do until morning, Amy and Bill returned to their cabin for a few hours' sleep. At dawn, Dian assembled search parties. They set off into the forest to find the remaining members of Uncle Bert's family.

One wounded and two dead was the immediate toll of the poachers' attack.

The clearing where Uncle Bert's body had lain the previous day was the obvious starting point. Flattened vegetation encircled the area. Several trails fanned out from there. One was the poachers'. Another was the gorillas'. A third seemed to be that of a solitary ape. Bill followed the lone gorilla's trail until he came across a dark mass, sprawled face down in a bed of thick plants.

The gorilla had a single gunshot wound in her back. When Bill called for help to roll her over, he discovered it was Macho, Kweli's mom. Months earlier, Macho had looked on as her two-year-old, Kweli, wandered over to Amy and touched the hem of her jeans. Now Macho was dead. Where was Kweli?

Amy had the answer to that question by day's end. She followed Group 4's path across the Rwandan border and into Congo, where she was relieved to find the group—especially Kweli. She had feared poachers might have captured the youngster to sell to an exotic pet collector. But Kweli, she noticed, was not his normal self; he was favoring one side of his body. An older gorilla picked at a bullet wound in Kweli's left shoulder. Amy wanted to treat Kweli's injury, but Karisoke had no means of capturing him.

Back at camp, Amy was charged with extracting the bullet from Uncle Bert's body in hopes of finding evidence that would lead to his killers. She hadn't dissected an animal since biology lab in college and didn't know where to start with Uncle Bert's headless corpse. Worse still, her only surgical instruments were a machete and a pocket knife.

Even though Uncle Bert's head was gone, Amy couldn't help feeling like she was violating his body. She sawed through his tough skin with the knife and felt around inside his chest cavity for a piece of metal that turned out not to be there. The bullet had passed straight through Uncle Bert's body.

Kweli's mother, Macho, lies face down in the undergrowth where Bill found her. She had been killed by a poacher's bullet in her back.

One wounded and two dead was the immediate toll of the poachers' attack. In the coming weeks, however, the casualties among the group of gorillas Dian Fossey had spent more than 10 years studying would rise to five: Silverbacks fighting for control of the family killed two infants, and Kweli succumbed to his injuries.

Dian was devastated. "They're all going to die," she said.

~ *Seeking Safety*

As the autumn of 1978 unfolded, Amy and Bill's research started to suggest otherwise. Bill's gorilla census showed the mountain gorilla population held roughly the same number of animals—about 260—as it had 5 years earlier. Amy's studies revealed that the gorillas still roamed through a lot of healthy habitat. If she and Bill could figure out a way to control poaching and protect habitat, Rwanda's mountain gorillas might just make it.

Defending that habitat would be tough. With Rwanda's human population growing rapidly, the government announced plans to convert 12,500 acres, a third of Volcanoes National Park, into cattle-grazing land. Rwanda's economy was very weak.

The cattle-raising project promised to bring in $70,000 of new revenue per year. That was a lot of money by Rwandan standards.

But such a move would reduce the gorillas' important bamboo feeding zone and chop the park into pieces. Since 1958, more than a third of the park had been cleared for farming. If another third disappeared, it was certain the gorillas eventually would, too.

It would be difficult to convince the government that a park full of gorillas was more valuable than the new cattle-grazing area. Annual entry fees to the park totaled just $7,000—one-tenth the income promised by the cattle project. Amy and Bill were desperate to find a way to protect the gorillas' habitat.

A Different Approach to Conservation

Like his wife, Bill Weber has always enjoyed nature. Not until he witnessed threats to Africa's wildlife during his stint in the Peace Corps, however, did Bill decide to pursue a career in wildlife conservation.

Though Bill did not have a biology background, he had taken psychology and political science classes that taught him how to work well with people—a handy trait to have for a career in conservation. Bill's experience teaching English and geography to teenagers in Zaire would also prove beneficial. "We got to know the people in a way that we could absorb their points of view and get a better understanding of the reasons for pressures facing wildlife," Bill says.

When Bill returned to the United States in 1975, he set out to find a graduate school that would allow him to study conservation without an emphasis on biology.

Bill and Amy searched the country for four months, camping in national parks. They both ended up at the University of Wisconsin-Madison—the alma mater of several pioneers in wildlife conservation, including John Muir, Aldo Leopold, and George Schaller.

Bill was working on his doctoral degree in land resource management at the university's Institute for Environmental Studies when he and Amy left for Rwanda. As Amy plunged into her ecology studies, Bill focused on how Rwandan land use practices were affecting gorillas.

Within weeks of arriving in Rwanda, Bill began a mountain gorilla census. He had important questions to answer: Of the 250 to 275 gorillas that had been counted by another researcher five years earlier, how many still existed? Had the rapid decline of the population, from 450 in the 1960s, continued?

When he finished the survey in the autumn of 1978, Bill knew that roughly 260 gorillas,

~ A Three-Pronged Protection Plan

After Digit's death in February, the British Flora and Fauna Preservation Society established the Mountain Gorilla Preservation Fund to raise money for gorilla conservation in the Virungas. Private donations to the fund reached roughly $100,000.

Inside their dimly lit cabin, Amy and Bill talked about how they would spend that money—if it was up to them—to help the mountain gorillas.

including lots of young animals, survived in the Virungas. That meant the population was stable, with the potential to grow. But Bill and Amy still needed to convince the Rwandans that gorillas—like the land they farmed—were a valuable resource.

Bill decided to start spreading this message among Rwanda's schoolchildren. Many of them would one day grow up to be politicians, government officials, and wildlife managers. Of Rwanda's 29 secondary schools, Bill visited 27. Most of the students had never seen—or even heard of—their country's mountain gorillas.

Bill showed films of the apes and explained that the animals were in danger. Unless protected from poaching and forest destruction, he told them, the gorillas would disappear forever. Many students expressed concern for the apes' welfare. They also talked with pride about how their small country held the key to the mountain gorilla's future. That national pride became a key part of Bill and Amy's campaign to save the mountain gorillas when they helped launch the Mountain Gorilla Project in 1979.

Faced with a cattle project that threatened the apes' future, they decided the best solution would be a gorilla conservation project with three essential features:

- anti-poaching teams to protect the gorillas from hunters,
- education to help Rwandans learn the value of their country's wildlife and natural resources, and
- well-monitored gorilla tourism to bring jobs and foreign money to the country.

After much discussion, Amy and Bill presented their proposal for a three-pronged Mountain Gorilla Project to the Rwandan government. "Can gorilla tourism really make more money than cattle?" the government officials asked.

A Rwandan girl learns about gorillas and other wildlife. Educating the people of Rwanda about their native plants and animals was an important part of the Mountain Gorilla Project.

Bill assured them that Rwanda could annually attract 3,000 tourists willing to pay $25 each to see gorillas in the wild. That would bring in $75,000—$5,000 more than the cattle project promised.

And that would be just the beginning. As time passed, more and more eco-tourists would come to Rwanda, happily paying big bucks to slog through freezing rain and mud for the once-in-a-lifetime experience of seeing gorillas in the wild.

Time would prove them right.

After fearing the worst,

Amy was thrilled

to see **Bill**

standing there

on his own two feet.

7

Close Calls

The tattered appearance of a young, long-haired couple in jeans and torn field boots raised eyebrows in New York City's John F. Kennedy Airport.

"Open the trunk," a customs agent demanded.

Amy complied.

"And what have we here?" the agent asked, fingering hundreds of small plastic bags filled with plants.

She quickly explained, "They're gorilla foods. Dried gorilla foods. I study what gorillas eat."

Amy showed the agent a handful of permits to prove that none of her samples were endangered species. If that didn't convince him she was a legitimate scientist, the bags of gorilla dung did.

The agent stared at the dung in disbelief. Then he laughed. Finally, he closed the trunk and waved Amy on her way.

As she and Bill moved beyond customs, they joined the masses of people jockeying through the busy airport. After living in Rwanda's mountains for nearly two years, the hustle and bustle of the big city was jarring. Saying goodbye to Group 5 had also been upsetting to Amy. She still had so many questions: How much longer could Beethoven lead the group? What would happen to Pablo? Would he settle into a new role within the group? Or would he leave his family and live a solitary life?

Bill Weber *(opposite)* holds hail that fell in the Virungas. Fieldwork in Rwanda held many perils for Bill and Amy. Tourists *(above)* enjoy the wild experience of watching gorillas up close.

~ A New Chapter

MOUNTAIN GORILLA PROJECT

Someone else would have to answer those questions. Amy needed to return to school in Madison to analyze the data from her research at Karisoke and finish some more coursework. Bill would head back to Rwanda four months ahead of Amy. After presenting their three-pronged conservation proposal to the Rwandan government, the couple had been offered financial support to launch the Mountain Gorilla Project. It was set to kick off in the summer of 1979. No longer would they have to scrimp and save just to buy rice and beans. The salary—$500 a month for each of them—would go a long way in Rwanda. Finally, they would be able to afford luxuries such as cheese and canned tuna.

Even better, they would be able to help ensure that the gorilla tourism program was done right, with minimal impact on the endangered apes. Bill had to hurry back to Rwanda to condition, or get the gorillas used to, tourists who would soon come to see them.

In their nine years together, Amy and Bill had spent days and even weeks apart, but never four months. They promised to reconnect in Rwanda for Christmas.

~ Bushwhacked!

After another semester of graduate classes and with her data analysis complete, Amy boarded a plane for Rwanda on December 21, 1979. On the last leg of the trip, she heard a group of people in a Belgian airport talking about their plans to visit Rwanda's gorillas.

"But we're not sure we'll be able to see them," someone said.

"Why not?" asked another traveler.

"One of the guides was attacked by the gorillas."

Amy jumped into the conversation. "Was it a Rwandan guide?" she asked.

"No, an American. A big man with a beard. I've seen him, but I don't know his name," one of them answered.

The description fit Bill.

Frantically, she asked: "How badly was he hurt?"

"They say he was bitten in the neck, but I don't know how bad it was."

Frightened about what might have happened to her husband, Amy boarded the plane to Kigali, the capital. After an excruciating 12 hours, she disembarked and eagerly searched for Bill's face. She was early. *Maybe he's not here yet,* she thought. *Or maybe he's paralyzed in a hospital bed.*

Amy started to feel foolish about the big red bow she had worn for Bill as a Christmas gift. Then she saw him in the crowd, bandages on his neck, his arm in a sling, and a big smile on his face. After fearing the worst, Amy was thrilled to see Bill standing there on his own two feet. She hugged him carefully—he had two broken ribs—and asked, "What in the world happened?"

Bill told them the policy was unsafe, for both the gorillas and the tourists.

Bill explained: It was all a case of mistaken identity.

~ Laying Down the Law of the Jungle

Bill had been guiding a group of tourists on the mountain two weeks earlier. It was a big group of 16 people, much bigger than he liked. But the Rwandan tourism board was pressuring Bill to take every paying customer to see the gorillas; he was not to turn anyone away. Bill told them the policy was unsafe, for both the gorillas and the tourists. But the tourism board wouldn't listen. So he reluctantly trudged up the mountain with the crowd of 16 Air France employees that day, following the trail of one of the two gorilla families Bill had conditioned to tourist visits.

Or so he thought. Turns out it wasn't one of the usual gorilla groups. It was Group 6—Brutus's clan.

Eco-tourists crowd close for the photograph of a lifetime—mountain gorillas foraging for food in their natural habitat.

Brutus was a 400-pound silverback who lived up to his burly name. Back in the days when Bill was still counting the gorillas on the Virungas slopes, the monstrous silverback had once raced directly at him, screaming *Wraaagh!* Although that charge had ended with Brutus merely storming off into the forest, this one did not. When Bill accidentally came upon Brutus, the ape put on a repeat performance—then sank his two-inch-long canine teeth into Bill's neck.

Later at a friend's house in Kigali, where they were staying overnight, Amy cringed when she saw that the bite wounds extended as far as an inch deep into Bill's flesh on either side of his spinal cord. Half an inch to the left or right and he might have been paralyzed for life. The deeper of the two wounds, Amy noticed, was infected. She carefully opened the gash, removed the tainted tissue, and bandaged it back up. It was a routine she would have to repeat several more times before the wound fully healed.

Bill's brush with Brutus might have been avoided if fewer tourists had been with him that day or if he had been given enough guides. The guides could have monitored the movements of gorilla families traveling in the area. Hoping to set up stricter rules and receive permission to recruit more guides to monitor the gorillas and detect poachers, Amy and Bill visited the director of Rwanda's tourism board, Benda Lema, in Kigali.

Rising from his desk, the director gave no sign that he noticed the bandages on Bill's neck. He shook Bill's hand, then Amy's, and invited them to sit down. Benda Lema told Amy he was glad she was there to "help her husband." Amy hid her annoyance at the remark. More important matters were at stake. She and Bill stressed that gorilla tourism could not continue safely without a limit on the number of tourists who could visit the gorillas each day.

A guide at Volcanoes National Park stands by an information board that provides details about the park and its trail.

Bill's bandages must have had an effect on the director after all. He canceled his previous policy of unlimited visitors and set the daily maximum of tourists to six. When Bill and Amy asked for more guards to monitor the area and ward off poachers, he agreed.

All in all, it had been a productive day of dealing with the Rwandan government.

~ *In the Shadow of Visoke*

Amy returned to a simple but satisfying life in Rwanda. Her previous home on the slope of Mount Visoke was replaced by a round metal structure at the volcano's base. It had a dirt floor, a cone-shaped roof that ricocheted rain like gunfire, and an outdoor latrine—a hole in the ground surrounded by four crude walls. Bill added a custom-made wooden seat as a Christmas gift.

Basic, too, were the hut's furnishings. There were two foam mattresses, a desk, a folding chair, and several big trunks that doubled as tables. Outside was a bench made of eucalyptus wood.

It provided the perfect perch from which to enjoy the morning chorus of bird songs and the breathtaking views of the eastern mountains of Sabyinyo, Gahinga, and Muhavura.

Amy had little time to ease into her new surroundings. She quickly began shepherding groups of gorilla tourists. By 8:00 most mornings, she could hear the first arrivals' cars climbing up the rocky road that ended in a small parking area beyond the hut. The local Rwandans who worked as gorilla guides, trackers, and porters were already waiting for the tourists, just as eager to connect with paying customers as the foreigners were to spot gorillas.

~ Stilgar's Charge

Only rarely was the six-person daily limit broken. One day in late January 1980 seven tourists showed up in Amy and Bill's driveway. Two of them were an elderly couple. Making an exception for the couple, Amy offered to take them for a leisurely gorilla hike.

Bill, now fully recovered from his broken ribs and punctured neck, guided the others at a faster pace. His group located the gorillas, Group 11, and sat watching them for about 45 minutes. Then a tracker arrived to tell him that Amy and the couple were close by.

Bill rounded up his group and backed away to allow Amy a chance to show the couple the scene. Amy waited a bit before approaching the gorillas, but her caution was in vain: Stilgar, the silverback of the group, had been spooked by the appearance of more people. He started strutting, as if preparing to charge.

With visions of Brutus dancing in her head, Amy dropped to her knees. "Get down!" she insisted.

"Wraaagh!" Stilgar bellowed. He ripped a tree sapling from the ground and charged, delivering a light blow to Amy's skull as he passed.

Breathless, Amy looked up to find Stilgar gone. Wobbling to her feet, she found the man lying on the ground, white as a ghost. The woman stood nearby, grinning from ear to ear. "I've never seen anything like this," she told Amy, beaming. "It's wonderful!"

Gorilla Speak

Amy saw firsthand from her encounter with Stilgar—and Bill from his with Brutus—how gorillas react when they're scared. When a silverback bellows *"Wraaagh!"* you can assume that's ape for "Buzz off, intruders!"

Though mountain gorillas are by nature shy, the 400-pound silverback is an aggressive leader and protector of his group.

He will charge at anything, or anyone, that threatens him or his family. The silverback may drum his chest in a rapid, hollow-sounding *pok, pok, pok* before he charges, a behavior known as chest beating.

Bellowing and chest-beating are not the only ways gorillas communicate. They also have a small repertoire of vocalizations. All gorillas in a group give soft "belch" vocalizations in a variety of situations. If the silverback is acting rowdy, a female sitting nearby might make a low rumbling sound as if to say, "I'm here, but I'm not going to bother you."

Young gorillas vocalize, too. When they wrestle and chase each other, they let out a series of "chuckles," similar to children giggling.

Another sign of gorilla contentment is "singing." Amy describes singing as a series of rumbles with different tones added to them. Sometimes, she would hear a mother and her infant sing a duet. The mother was a baritone, while the baby was more of a soprano. "When gorillas 'sing,'" Amy says, "it can be like us humming in the shower. They're saying, 'Hey, life is great!'"

Amy wanted to tell the woman that her first encounter with a gorilla had almost been her last. Instead, she smiled and explained that it was not a good time to pursue the gorillas; they had better head back. Amy never broke the "rule of six" again.

Nyungwe was
the richest **forest**

Amy had
ever seen.

8

Going Beyond Gorillas

More than 1,000 paying tourists visited Rwanda's gorillas in 1980, the first full year of the Mountain Gorilla Project. That was double Volcanoes National Park's attendance the year before. The number of anti-poaching guards in the park force nearly tripled, from 14 to 40. And park revenue quadrupled. The plan to save Rwanda's mountain gorillas was under way and gaining support.

Amy's son, Noah Gerhardt Weber *(above)*, was born in Madison, Wisconsin, on November 1, 1980. Motherhood did not slow Amy down. She still enjoyed the adventures of field research *(opposite)*.

With the Virunga cattle project no longer a threat—and with poaching way down and gorilla numbers on the rise—Amy and Bill concluded that the gorillas were doing much better than when the couple had arrived two years earlier. It was time to move on. Other wild places in Rwanda were in danger. And few of them were home to such charismatic creatures as mountain gorillas, making efforts to protect them much more challenging.

Amy wanted to conquer those challenges, but she and Bill agreed that they needed to finish their postgraduate degrees back in the United States. However, before leaving, there was one more thing Amy wanted to do.

~ *The View from the Top*

Amy wanted to take in the view from the tops of the five volcanoes that form the country's backbone. Aside from the pure enjoyment of hiking the volcanoes, Amy wanted to map a network of backpacking trails across Volcanoes National Park. A system of hiking trails would offer another lure for tourists and a little extra insurance for the gorillas' future.

That final summer Amy began her journey across the Virungas on the eastern flank of Mount Muhavura, at Rwanda's border with Uganda. Accompanied by Rwandan guides and three French teachers, she hiked across slate-gray lava flows covered by sage-colored moss and spotted with bright red flowers called red-hot pokers. She climbed all the way to Muhavura's summit. There she could see far to the north, where Uganda's Bwindi Impenetrable Forest holds the only other population of mountain gorillas in the world.

In the summer of 1980, Amy climbed to the second-highest point of Sabyinyo, this 12,060-foot-high volcano. The name means "teeth of the old man" in Swahili.

Masses of colorful plants called red-hot pokers cover Mount Muhavura.

Hundreds of years ago, Bwindi's gorillas mingled with the Virunga population. But farmers and herders cleared the land between the parks, marooning the gorillas on two separate forest islands in the mountains.

After a night in Muhavura's bamboo valley, Amy climbed Gahinga, a jagged volcano pocked with craters and carpeted by pink orchids.

Next up was a frightening climb on Sabyinyo. Amy crawled hand over hand up a near-vertical ridge of rock to the top of Sabyinyo's second-highest peak. Then dark clouds rolled in, forcing Amy and the three other climbers to retreat. A day later, she gazed down from the summit of Visoke, her eyes following an elephant trail along the round crater's slope to a deep caldron of blue water.

Karisimbi—Africa's fourth-highest mountain and the last summit on her trek—provided the coldest climb. In the thin, icy air near the summit, Amy found it tough to breathe. Still, she mustered enough energy to goad her three companions into a giant snowball battle before declaring an end to their adventure and heading back.

~ *Born in the U.S.A.*

A month after her adrenaline-fueled trek across the Virungas, Amy returned to the quiet life of a graduate student in Madison and the luxuries of American living. She enjoyed direct-dial telephone calls, drinkable water straight from the kitchen faucet, and flush toilets. She also liked washing machines, fully stocked grocery stores, daily newspapers delivered to the doorstep, and shopping—

Proud parents Amy and Bill enjoy time with their son Noah in 1981.

baby shopping. Amy was five months pregnant with the baby she and Bill had decided to have shortly after Bill's near-death encounter with Brutus.

When he was born just a few weeks before Thanksgiving in 1980, Noah Gerhardt Weber had already covered more terrain from the comfort of Amy's womb than many people do in their entire lives. After he was born, Amy and Bill juggled their graduate class schedules so that one of them would always be with Noah. And, when necessary, Noah went along to classes. In December 1982, Amy received her master's degree in zoology.

~ *Into Africa*

By the time Noah was two years old, Amy had to get a little more creative with child care; Bill had begun making trips to Africa on a regular basis. He was now a consultant for the U.S. National Park Service and the U.S. Agency for International Development (USAID). For his job he traveled through East Africa documenting threats to wildlife and protected areas.

During one of his trips, the USAID director of Burundi, Rwanda's southern neighbor, asked Bill to help the government write a management plan for a small mountain forest in the southern part of the country. Accepting the job and dividing the work in two, Bill and Amy returned to Burundi in 1983 with three-year-old Noah in tow.

Accompanying them on the trip to help take care of Noah was a fellow graduate student, Bette Loiselle. She had been awarded what Amy called the "Vedder-Weber doctoral nanny scholarship."

Together, the four travelers made a good team. Amy studied the forest's mammals while Bill looked into social and economic issues in the surrounding villages, Bette focused on ornithology, the study of birds, and Noah took care of entomology, the study of insects. The toddler was fascinated by bugs of all types, especially those he could pick up for closer inspection. Noah quickly learned not to touch the African caterpillars that stung his curious little fingers.

~ *Monkey Haven*

During their time in Burundi, Amy and Bill returned to Rwanda to see a place they'd heard about for years: the Nyungwe (n-YOONG-gway) Forest. Some conservationists believed this 400-square-mile rain forest was the crown jewel of Rwanda's natural areas. Yet little was known about it.

Nyungwe was the richest forest Amy had ever seen. The canopy, the umbrella-like upper level of branches and foliage, was made up of giant trees of all shapes and kinds—roughly 200 species in total. (The Virungas, by comparison, supported only about a dozen tree types.) In Nyungwe, 200-foot-tall mahoganies towered over

Amy takes three-year-old Noah for a walk through a mountain forest in Burundi *(left)*. These explorations gave Noah close-up views of praying mantises *(above)*, African caterpillars, and many other insects.

The Nyungwe rain forest in Rwanda is home to 13 species of primates, including black-and-white colobus monkeys *(right)*, mountain monkeys, owl-faced monkeys, and chimpanzees.

the landscape, dwarfing the 100-foot Hagenia trees familiar to Amy from the Virungas.

Birds were everywhere, too, as Nyungwe is home to nearly 300 species. As Amy hiked through the forest, she saw iridescent green, red, and yellow sunbirds drinking nectar from canopy flowers. Large birds called turacos flashed crimson-red feathers as they took wing.

Amy wasn't surprised to see crowned eagles soaring overhead: The large, high-flying predators liked to make meals out of monkeys, snatching them in their razor-sharp talons. Nyungwe was jam-packed with primates—13 types, including baboons, mischievous vervet monkeys, two species of small nocturnal primates called bush babies, several hundred chimpanzees, and majestic-looking black-and-white colobus monkeys. Amy could easily see dividing her time between Nyungwe and the Virungas.

~ And Baby Makes Four

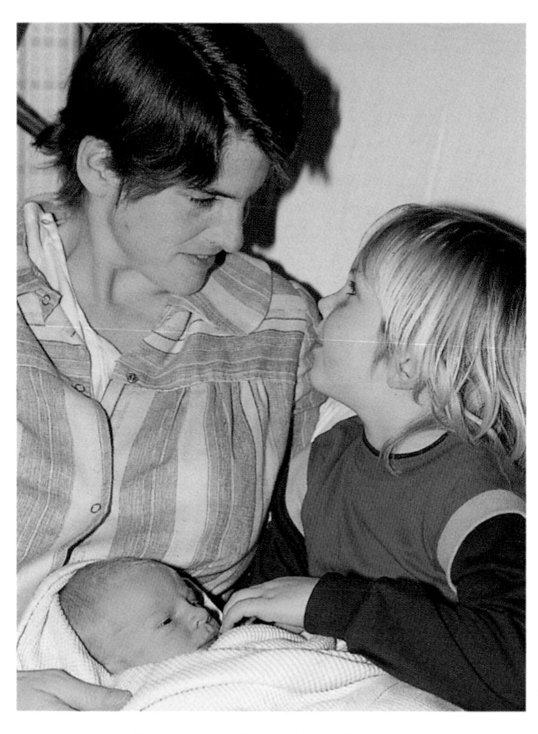

Noah's little brother Ethan was destined to become Noah's partner on many an African adventure.

By early 1985, Amy had enough funding from the Wildlife Conservation Society for a 14-month biological survey of Nyungwe that would mesh nicely with Bill's plans to work on a resource management project near the Virungas. The timing of Bill's full-time position was fortunate. They were nearly finished with their degrees and would need the steady income to support their growing family. On March 1, 1985, their second child—Ethan Heller Vedder—was born in Madison.

Only four months later, Amy boarded a plane to Rwanda, where Bill had returned to work a month earlier, with baby Ethan in her arms and little Noah by her side. Packed in the boys' suitcases were enough clothes for three years of growth.

The thought
that a ***killer*** *could still*

be lurking outside
was ***unsettling***.

9

Murder on the Mountain

"Have you heard the news?" the receptionist asked.
"What news?" Amy answered.
"Mademoiselle is dead."
"Mademoiselle Fossey?"
"Oui."

World-famous gorilla researcher Dian Fossey was murdered inside her cabin *(opposite)* at Karisoke on December 26, 1985. She had spent almost 20 years amid the beauty of places like Mount Visoke *(above).*

Amy and Bill stood at the reception desk in shock. They had stopped by the national parks office in Kigali that morning only to arrange the permits required for a WCS colleague coming to see Amy's early research in Nyungwe. Now they were faced with the news that Dian was dead.

A passing official spotted Amy and Bill and confirmed the report. Yes, he had received a radio message saying Dian Fossey was found dead in her cabin, but there were no further details.

Soon they were joined by one of the top U.S. authorities in Kigali, Kathleen Austen. Kathleen said a Rwandan team was on its way to Karisoke to investigate and she was going to meet them there. Knowing that Amy had worked with Dian at Karisoke, Kathleen asked Amy to come along. Bill could watch Noah and Ethan for the night. Amy reluctantly agreed.

Hours later, stepping off the trail that spiraled up Visoke, Amy and Kathleen covered the last steps to Dian's cabin just before nightfall. Inside, an inspector wearing a trench coat was interrogating Dian's housekeeper. Another man was taking notes on a small portable typewriter. The investigator paused to greet Amy and Kathleen, then showed them to the bedroom.

Dian must have grabbed for her gun; it was on the floor, near her hand.

Amy gulped as she took in the gruesome scene. Dian lay sprawled out in her long underwear on a floor mat beside the bed. The wounds on her face and neck, and the blood-spattered machete on the floor, told the story: Dian had been brutally murdered.

~ *In Search of Clues*

The inspector moved across the room and pointed to a gaping hole that had been cut in the corrugated-tin wall of Dian's bedroom. This was apparently where her assailant had entered, he explained.

Amy wondered: *Why hadn't Dian escaped when she heard someone sawing through her bedroom wall?*

Maybe she had tried. Dian's bed appeared slept in, but there were obvious signs of a struggle: Books and papers were strewn across the floor. Dresser drawers hung open, some of their contents spilling out. Dian must have grabbed for her gun; it was on the floor, near her hand.

The Rwandan investigators lifted a few fingerprints from the top of Dian's dresser. They took photographs of footprints on the path outside the cabin and followed Kathleen's suggestion to remove the hairs clutched in Dian's fist.

Later that night Amy lay awake in bed at the guesthouse near Dian's cabin. The thought that a killer could still be lurking outside was unsettling. In her 13 years as director of primate research at Karisoke, Dian Fossey had made plenty of enemies—even among her own staff.

Others had experienced the type of treatment Amy and Bill had endured from Dian in the weeks after Mweza's death. Still,

Amy found it difficult to imagine that anyone among the Karisoke staff was capable of murder. She also thought about the overwhelming experience of seeing Dian's dead body. Amy knew before making the climb to Karisoke that Dian was definitely dead, but she still had not been prepared to see her corpse. And even though Amy's relationship with Dian had been strained at times, she had gotten to know the woman behind the legend fairly well while working at Karisoke.

~ *A Complicated Personality*

Amy remembered how much Dian had liked holidays. One Halloween, Amy carved a face with a toothy grin in a hollow gourd, stuck a candle inside, and propped the jack-o'-lantern on Dian's doorstep. After knocking on the door, Amy ran and hid in some bushes nearby.

Nothing happened.

Amy knocked again. This time Dian shouted, "Who is it?"

Should I knock a third time? Amy wondered.

Just then the door flew open and there stood Dian, waving a pistol. When she noticed Amy's gift at her feet, she let out a giggle, then picked up the "pumpkin" and carried it inside.

Months later Amy repeated the ritual on Easter, when she left a few hand-painted eggs in a nest of galium, the plant that gorillas eat most frequently, on Dian's doorstep. Recalling the pistol in Dian's hand at Halloween, Amy didn't wait around to watch the woman's reaction. The next day Amy learned that Dian had thanked a newly arrived graduate student for the thoughtful Easter surprise.

Dian Fossey viewed the gorillas she studied as individuals. She named them and shared their life stories with a worldwide audience.

Nonetheless, Amy continued to reach out to Dian. On Christmas Day 1980, Amy was tracking Group 5 when she realized the gorillas were moving closer and closer to camp. Dian was only 48 years old at the time, but her health was failing, and hiking far to see the gorillas had become too much of a struggle. *Yet,* Amy thought, *this might be Dian's chance to see them.* The gorillas were going to be only a 10-minute walk from her cabin. She raced back to camp and told Dian the good news: The gorillas "brought you a present. They're waiting only 200 yards away!"

"No, I can't," Dian replied.

~ *Buried in the Gorilla Graveyard*

Dian Fossey was buried beside her old friend Digit in the gorilla graveyard of Karisoke Research Center.

Five Christmas holidays had passed since Dian uttered those sad words. It had been the last Christmas Day Amy spent at Karisoke. Now, as she joined the small funeral procession winding its way up Mount Visoke on New Year's Eve 1985, Amy thought back to the bitterness in Dian's voice when she refused to go see the gorillas. On Dian's casket, Amy placed a bouquet of favorite gorilla foods: wild celery, galium, blackberries, and thistles. Then she watched the casket descend into the ground beside Digit's grave.

~ The Accused

In the weeks after the murder, journalists flocked to Rwanda from around the world. All of them wanted to know who killed Dian Fossey—and why?

The answers to those questions eluded the Rwandan authorities. They were pondering two possibilities: Gorilla poachers had killed Dian in revenge or she had been slain by a disgruntled Karisoke researcher.

The crime scene evidence did not point clearly to a culprit. But the presence of journalists and persistent questions from the American embassy in Rwanda were creating pressure: Someone had to be found responsible for the crime. After several months, the Rwandan Ministry of Justice announced that it would charge with murder an American researcher who had arrived at Karisoke a few months before Dian's death. Also charged was Amy's old friend, the gorilla tracker Rwelekana.

By the time the American researcher was charged with the murder, he had fled the country. And despite Amy's efforts to give the authorities evidence proving that Rwelekana lacked the motive or the opportunity to commit the murder, her friend was jailed.

Several months later, Rwelekana was killed by his captors. His cause of death was falsely listed as "suicide." As the authorities closed the books on Dian's murder, Amy quietly mourned Rwelekana's. To this day the identity of Dian Fossey's murderer remains unknown.

Occasionally, Amy and Bill
took Noah and Ethan

out of town
on weekend trips.

10

Family Life in a Foreign Land

Amy had gotten used to shivering beneath the covers in the tin shack at Karisoke in 1978. In 1980 she had adapted to the boiling daytime heat in her second metal hut at the base of Visoke. But when she and Bill returned to Rwanda with their two growing boys in 1985, they picked a home that would offer a bit more for the family. They found a large one-story colonial house in the town of Ruhengeri.

Amy's life in Rwanda in the mid-1980s focused on the Four F's: family, flora, fauna, and fun. On the opposite page, she carries Ethan across a bamboo bridge in eastern Congo. Above, Bill and Noah explore Mount Sabyinyo together.

Shortly after getting settled, Amy began searching for a nanny who could help with the boys while she and Bill were busy with fieldwork. One of the women she interviewed was Clementine Uwimana, a young Rwandan. As she chatted with Amy, Clementine picked up Ethan and held him close. Watching her baby return the woman's warm embrace, Amy knew the job was meant for Clementine.

~ *Mom's Forest Office*

During the week, Clementine shuttled Noah back and forth to the French kindergarten in Ruhengeri and looked after Ethan during the day. On weekends she often accompanied Amy, Bill, Noah, and Ethan on research trips to the Nyungwe Forest.

Amy and Bill rented this house in the Rwandan town of Ruhengeri. The gardens outside overflowed with flowering shrubs and lemon, papaya, and pomegranate trees.

Clementine became good at spotting monkeys. Speaking in a whisper, Amy explained what the curious creatures were doing. The mountain monkeys—large black animals with rusty-colored backs and white beards—traveled mostly on the ground, their tails held high like question marks. Blue monkeys announced their location with loud *piao* calls that echoed through the treetops. Red-tailed monkeys, showing off the trademark white spot on their noses, peered down from branches on the forest's edge.

Every once in a while, Amy pointed out a mona monkey. Against the mona's slate-gray body, the creamy white fur on its belly and the inside of its arms and legs made the creature look like it was wearing a luxurious leotard. There were also chocolate-brown mangabeys, which Amy referred to as the punk rockers of the forest. They had fuzzy cloaks of fur, dramatic cheekbones, and spiky hair standing straight up on their heads.

Noah and his friend Gael van de Weghe *(below, left)* probe the soil of the Nyungwe Forest in search of earthworms and ants.

Sometimes, Noah's five-year-old friend Gael van de Weghe came along for a weekend of observing the monkeys in Nyungwe. Running out in front of Amy, Noah and Gael combed the forest floor for foot-long earthworms and pretended to have sword fights with sticks.

One day when Amy spotted a tree called a strangler fig, she called to the boys to come take a look. The strangler fig, she explained, was engaged in a fierce battle, just like the boys with their swords. Noah and Gael stared at Amy with wide eyes as she explained that strangler figs grow on top of other trees, choking their hosts, blocking their sunlight, and winning the game of life. Suddenly, Amy was speaking the boys' language.

~ *Weekend Getaways*

Occasionally, Amy and Bill took Noah and Ethan out of town on weekend trips. One afternoon they returned to the Virungas to hike up the lower part of the trail that climbs Mount Sabyinyo. Baby Ethan rode along in a backpack while Noah leapfrogged up the mountain path, jumping from one water-filled elephant footprint to the next.

On another weekend Bill took Noah to Rwanda's Akagera National Park while Amy and Ethan stayed in Kigali. After a full day in the park, Bill was standing at the main lodge's check-in counter when he suddenly noticed that Noah was missing. He ran back toward the car, searching frantically for his son. An overwhelming panic washed over Bill—he had a feeling that Noah had gone to see the lodge's caged chimpanzees. Passing the apes on their way in, Bill had warned Noah not to go too close—the chimps were acting aggressively.

Fearing the worst, Bill raced toward the chimp cage.

Ethan *(left)* and Noah enjoy the unique view from their parents' shoulders.

He arrived just as a long, hairy arm reached past Noah's extended handful of leaves and pulled the boy's arm through the bars. The other chimps joined the attack, screaming and biting.

Beating back the chimps, Bill pulled Noah to safety. He used the only supplies he could find—a washcloth and some ice—to stop the blood gushing from Noah's arm.

People at the lodge told Bill he could find medical help at a Red Cross camp roughly 10 miles north of the park. A Sudanese doctor there cleaned out Noah's 20 bite wounds with alcohol and gauze. Then Bill raced Noah back to Kigali, where their pajama-clad family doctor examined the injuries at 10:00 in the evening. The doctor told Noah he was a brave little boy and gave him some medicine to prevent infection and numb the pain.

Not long after Noah had been attacked in Akagera, he was walking with Amy in Nyungwe when they spotted some chimps. Concerned that Noah might be anxious, Amy asked him what he thought of seeing the animals again.

"Nice," Noah said.

"You're not afraid?" Amy asked.

"Nope," Noah answered. "Those chimps in Akagera were crazy from living in cages. These chimps are happy 'cause they're in their own homes."

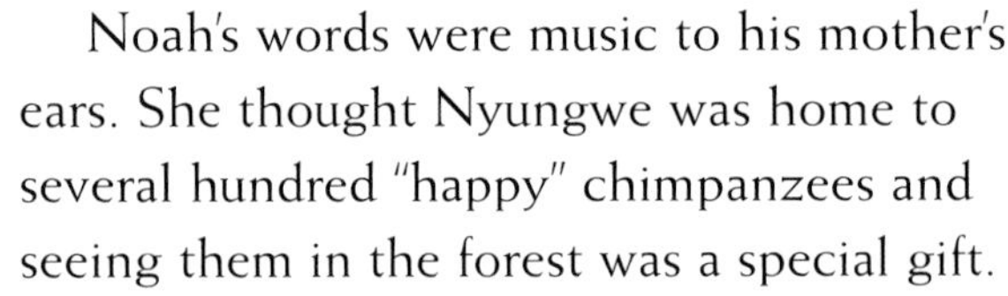

On one of the family's visits to Group 5, Cantsbee, this curious young gorilla, tried to untie Noah's shoelace.

Noah's words were music to his mother's ears. She thought Nyungwe was home to several hundred "happy" chimpanzees and seeing them in the forest was a special gift.

Before long, Amy and Bill took Noah to meet Amy's extended family, gorilla Group 5. They had wanted to take Ethan, too, but they knew it would be dangerous to put the rambunctious toddler anywhere near a clan of big, powerful apes. Even Noah was a little afraid of the gorillas, especially when Pablo pretended he was about to charge. *Typical Pablo,* Amy thought. *Always creating a ruckus.*

Then seven-year-old Cantsbee tried to untie Noah's bootlaces. "Don't worry, Noah," Amy assured him. "He's just trying to play." But when Noah still looked concerned, Bill stood over his son on all fours, just like a silverback protecting his young.

~Magical Monkeys

As the months rolled by, Amy began to uncover some of Nyungwe's secrets. Before returning to Rwanda, she had heard rumors from other scientists about enormous numbers of colobus monkeys living in the forest. Most colobus monkeys are black and white, but Nyungwe's colobus are unique. Their jet-black coats are offset by long white hair that drapes regally from their shoulders. The serious expression on their faces looks strange with two clownlike tufts of white hair jutting from their cheeks. And as the animals leap from one treetop to the next, you can spot the bushy white plume on their tails trailing in the wind.

Colobus monkeys swarm the branches of a tree in the Nyungwe Forest. During one of her wildlife surveys in the forest, Amy counted 353 colobus monkeys in a single group.

That was the view Amy had seen for months while tracking the monkeys' movements, making it impossible for her to figure out how many animals were in the group. Finally, she had a break one day when she realized the monkeys were headed toward a large gap in the canopy. They would have to come down to the ground to cross a road, providing the perfect opportunity for Amy to do a head count.

She raced through the forest, determined to get to the road in time. With her heart pounding, Amy reached the road just before the monkeys started piling into the trees at its edge. Breathlessly, she waited.

Finally, one of the monkeys took the first leap. One by one, the others followed. Amy watched and counted as a torrent of black bodies with streaming white capes and tails jumped to the ground and filed across: 1, 2, 3, 10, 20, 30, 100, 200, 300. . . . It took two and a half hours for all 353 colobus monkeys in the group to cross. Never in her wildest dreams had Amy expected to see so many! They confirmed her sense that Nyungwe was a magical place brimming with unusual wildlife.

Nyungwe Plants and Animals

The Nyungwe Forest is one of the largest mountain rain forests remaining in Africa. Its varied ecosystems contain a wide range of plants and wildlife, including 190 species of trees, 100 types of orchids, 275 species of birds, 13 types of primates, and many other mammals, including leopards, elephants, forest pigs, and antelopes. And there are probably even more species yet to be identified.

A fringed orchid.

One of several forest species of impatiens, each with different colored flowers.

The *Symphonia globulifera* tree. Its fruit and seeds provide food for some of the forest monkeys.

A wild gladiolus flower.

A multicolored caterpillar, species yet to be identified by biologists.

A three-horned chameleon with eyes that pivot in all directions.

A mountain monkey with its tail in the typical "question mark" position.

A red butterfly, one of more than 100 species in the forest.

~ Strange Twist

In the spring of 1987, Amy and Bill returned to the United States for a brief visit to the Wildlife Conservation Society (WCS). The organization had paid for Amy's research in Nyungwe, and she needed to let them know how the work was going. At the end of her presentation, staff members at WCS applauded and shook Amy's hand, congratulating her on a job well done.

Then they offered Bill a job. They wanted him to be deputy director of the organization's international program. Bill deserved the position, Amy reasoned. After all, he had been working on something unique: the social and economic side of conservation, which often got overlooked. He studied the way development, land use, and public perceptions affect wildlife. Amy asked the staff if they could hire her as well. They replied that it might not be beneficial to have a married couple working so closely together in such a small department. Though Amy was disappointed, she was happy for Bill. He had always supported her work, and this time it was her turn to watch his career take off.

Though Amy was disappointed, she was happy for Bill. He had always supported her work, and this time it was her turn to watch his career take off.

~ American Treats

Before their departure from Rwanda, Amy and Bill moved to a small cottage on a tea plantation just east of Nyungwe. From there Amy had easy access to the forest as she finished her survey work. Bill could take long walks with the boys and ride bikes along the roads that crisscrossed the plantation. During these final months in Rwanda, Noah and Ethan became best friends.

Most days the boys tagged along behind Alexi and Justin, two Rwandan men who worked on the plantation. The men were good sports about entertaining their pint-size companions.

They put Noah and Ethan to work—hoeing the garden, carrying buckets of fruits and vegetables, chopping firewood.

Meanwhile, Amy and Bill concentrated on helping Noah prepare to attend an English-speaking school for the first time when he returned to the United States. Noah had gone to a French kindergarten in Ruhengeri. The next year he had spent six days a week attending first grade in a Belgian school in Kigali. For his last six months in Rwanda, Noah's only teachers were his parents, who focused on helping him learn English and apply his math skills. They made the lessons fun and interesting. One assignment, for example, required Noah to count how many times he could jump rope without making a mistake. After each attempt, he wrote down the number. Then he practiced calculating averages and making graphs.

Amy and Bill also introduced the boys to traditional American holidays. For Halloween, they carved faces in pumpkinlike gourds. For Christmas, Amy decorated the cabin with homemade ornaments, branches cut from a pine tree, and a red cloth ribbon that she hung over the fireplace. Gifts for the boys were a challenge. Few toys could be bought in local markets. So that final Christmas, Amy and Bill bought a big bunch of balloons and spent hours blowing them up to the perfect size for volleying with the kids on Christmas morning.

Ethan *(top)* and Noah delight in some unaccustomed treats on their last Christmas morning in Africa—balloons, books in English, and imported apples and oranges.

~ *Homeward Bound*

Weeks later, Amy said a sad goodbye to Bill and the boys at the airport as they boarded a flight to the United States. She then returned to Nyungwe to finish her survey work and training for the Rwandans who were taking over. After dropping Noah and Ethan at their grandparents' house in Florida, Bill headed to New York to begin his new job—and to find a house for the family somewhere within a 20-mile radius of the Bronx Zoo.

Amy was glad that Noah and Ethan would finally get to know their grandparents better. Yet she felt guilty about staying behind in Rwanda—especially when her mother told her that Noah was unhappy and having trouble in school.

Not until a conference with Noah's teacher did Noah's grandmother, Marion, learn what was going on. The teacher told her that Noah was telling wild stories to the class: "He says he lived in Africa and spent time in the jungle."

"Well, he did," Marion replied.

"He also says he was bitten by chimpanzees," said the teacher. "The kids really had trouble believing that one."

"Well, it's true," Marion assured her. "Just look at the scars on his arm."

Ethan had some rough days, too. He got sick with the flu during the short stretch of time that he and Noah lived with their grandparents. When Marion cuddled him and tried to make him feel comfortable, Ethan told her, "You know, Grandma, you're my best friend." That story really touched Amy.

> Amy was glad that Noah and Ethan would finally get to know their grandparents better.

In March 1988, Amy was finally reunited with Bill and the kids in New York, where they moved into a house on a quiet street just a stone's throw from the high school where Noah and Ethan would eventually graduate.

Done with all her graduate classes, Amy had one more thing to do to get her Ph.D. She had to present her dissertation (a long research paper) to a group of professors at the University of

Wisconsin and defend her work. She finished that last important step and received her Ph.D. in zoology in 1989.

While Bill settled into his new job at the zoo, Amy began teaching ecology classes to college students at nearby Fordham University. Amy's university teaching career didn't last very long, however. Less than a year later, the zoo staff changed their minds about the pitfalls of having a married couple work together. They offered Amy a job working side by side with Bill, focusing on big development agencies and their impacts on wildlife conservation around the world. Some of her work took her to Rwanda, allowing Amy to return to her old stomping grounds on business.

Fortunately, she was nowhere near Rwanda in 1994 when the killing began.

Many of Amy's
friends and colleagues

were **murdered** or **missing**.

11

Guerrilla Wars

They looked like plaster mannequins. But they weren't. They were bodies. Unearthed from mass graves and covered with a chalk-colored preservative, the bodies lay spread out on tables in 78 rooms in the Murambi school. They were a stark and shocking testament of ethnic cleansing in Rwanda. Ethnic cleansing, or genocide, is the execution of people because of their race or religious beliefs.

Multiple crosses *(opposite)* mark a mass grave in Rwanda, where 800,000 people were murdered during the 1994 genocide. One victim's home is pictured above.

In life the bodies displayed in the school had been mothers, fathers, and children. In death they were now part of a new national memorial to the more than 800,000 people brutally murdered during a three-month bloodbath in 1994.

The sights and smells of the preserved skeletons made Amy feel sick as she walked slowly through more than a dozen rooms in the complex. She was seen the horrifying death toll firsthand. The Murambi school was supposed to have been a safe place for these people to hide. Now it was a grisly reminder of all the victims whose lives had been cut short simply because they were Tutsi, not Hutu—the two major ethnic groups in Rwanda.

American newspapers have often described the Rwandan genocide as the climax of a war between the Tutsi and Hutu "tribes."

But that paints the wrong picture. The truth is far more complicated. The groups share a culture, a language, and a religion, and they have coexisted for centuries. The Hutu were the first people to settle in Rwanda. They were traditionally farmers with sturdy builds. The Tutsi—generally tall, thin pastoralists who raised livestock—arrived later.

Major troubles between the Hutu and Tutsi surfaced at the end of World War I, when Germany surrendered control of Rwanda to Belgium. For decades the Belgians ruled the country indirectly through the Tutsi, who were given better opportunities in commerce and education than the Hutu. But the Belgian favoritism of the Tutsi angered the Hutu.

In the 1950s the Hutu began demanding better treatment. Such demands grew into calls for national independence. A Hutu uprising in 1959 killed Tutsi by the tens of thousands. Vowing to return one day to their native Rwanda, hundreds of thousands more Tutsi fled to the neighboring countries of Burundi, Uganda, and Congo.

~ *Guerrilla Fighters*

On October 1, 1990—mere months after Amy had joined Bill and the kids in New York—several hundred Tutsi men dressed in soldiers' uniforms stormed the Rwandan border. Known as the Rwandan Patriotic Front, or RPF, the men set out to take back the country they felt belonged to them.

On the first day, the RPF penetrated 40 miles into Rwanda and seized the headquarters of Akagera National Park, where Noah had survived that terrifying encounter with the caged chimps. On the second day, the RPF's field leader was killed and its offensive stalled. Many RPF guerrillas retreated to Uganda again.

Thousands of miles away, Amy watched these events unfold on the news and worried about her Rwandan friends and colleagues, as well as the gorillas and the park staff charged with protecting them. She hoped they would all be safe.

Three months later, in 1991, the RPF attacked again. This time the RPF captured weapons and ammunition, vehicles, and gasoline in the Rwandan city of Ruhengeri, where Amy had lived with her family. They also freed some 1,000 Tutsi prisoners from the Ruhengeri prison before retreating through the Virunga forest. After the attack, many Hutu grew suspicious of their Tutsi neighbors. In a few short days the Hutu killed 300 Tutsi.

~ *The Trigger*

Before hostilities broke out, Amy and Bill had planned to live in Rwanda each summer, when Ethan and Noah were not in school. They canceled those plans when the conflict started, but Amy still visited Rwanda regularly for conservation work. After a five-week trip in June 1991, she was growing more hopeful about the state of the country. The political troubles of the last year seemed to be quieting down. Tentative peace negotiations were taking place.

By the time she returned for a 23-day trip in the summer of 1993, however, the situation had worsened again. Earlier that year, the RPF had launched another major attack, taking over the entire Virunga region and leaving the fate of the gorillas up in the air. In the capital city of Kigali, the sound of gunshots and grenades pierced the night. Political demonstrations often ended in bloodshed. Amy managed to check in on her colleagues in Nyungwe, but travel around the country was limited—and increasingly dangerous.

Nearly a year later, on April 6, 1994, a jet carrying Rwanda's Hutu President Juvenal Habyarimana [hah-bee-yahr-ee-MAH-na] was blown out of the sky by two missiles launched from the ground. Within hours of his assassination, a nationwide killing campaign swept across Rwanda. One murder occurred every 10 seconds for 100 days, adding up to 800,000 dead Tutsi.

Within hours of his assassination, a nationwide killing campaign swept across Rwanda. One murder occurred every 10 seconds for 100 days, adding up to 800,000 dead Tutsi.

Although many Americans were unaware of the ethnic cleansing taking place in Rwanda during the spring of 1994, Amy and her family were plunged into a state of shock and dread. "Is Justin and everybody else on the tea plantation okay?" Noah asked. "What about my friend Gael? What about our nanny Clementine?"

Amy didn't know what to say. She had no idea how any of their friends and colleagues could survive this nightmare. She rebuffed journalists who asked her how the killing might be affecting the gorillas. It seemed wrong to discuss animal conservation when so many humans were dying.

All Amy could do was methodically clip newspaper articles, even though they rarely depicted the region's complicated history accurately. She watched in despair as CNN showed video footage of a flood of bodies washing down the Akagera River and into Lake Victoria.

~ *No Safe Harbor*

Tour guide Emmanuel Murangira displays the head wound he suffered during the three-day massacre of 27,000 fellow Tutsi at the Murambi school. Murangira was one of the few survivors; 48 members of his extended family were killed.

When Amy returned to Rwanda in 1999, the bodies in the Murambi school complex and the bullet wound in her tour guide's forehead underscored the horror of the genocide, one of the largest in world history. The guide, Emmanuel Murangira [moo-rahn-GEER-ah], showed Amy around the newly opened memorial. He told her how armed men had promised him and tens of thousands of other Tutsi protection at the school. But that was a trick. The armed men were there to make sure no one escaped.

Rwandan government soldiers arrived and opened fire on the people with rifles and grenades. Before long, local Hutu men joined in, hacking and stabbing with machetes and spears.

The gruesome killing lasted three days. Only a handful of people survived, and Emmanuel was one of them. Bleeding from a head wound and left for dead, he crawled away from the school at night and hid in a nearby stand of trees until the killing ended. Emmanuel was lying there, helpless, as the dead and wounded were tossed into massive trenches and buried. Among them were 48 members of his extended family.

The remains of 3,000 genocide victims were displayed in 78 rooms within the Murambi school complex.

The remains of 3,000 people—including Emmanuel's mother, his father, and his brother—were dug up and put on display when the school became a memorial to the 27,000 people killed at Murambi. The bodies represented only a fraction of the hundreds of thousands of Tutsi slain during 100 days of genocide in 1994.

~ *The Aftermath of Massacre*

In just a few months, Rwanda's population dropped from 7.6 million to 4.8 million, the same number it had been when Amy first arrived at Karisoke 16 years before. Many of Amy's friends and colleagues were murdered or missing.

In the years that followed, a very different picture emerged concerning the Virunga gorillas. After the outbreak of hostilities in 1990, combatants on both sides of the conflict had vowed to do everything they could to protect the mountain gorillas. Their respect for the apes was tied directly to their concern for the country's economy. Gorillas had become Rwanda's third-largest source of revenue (behind coffee and tea).

When the smoke cleared, it turned out that the rebels had kept their word: Only one mountain gorilla was killed during the Rwandan civil war and the genocide that followed.

"Don't forget *why* we do this work. Be sure to make time to *get out* into the field,"

Bill advised her. So Amy returned to *Africa* whenever she could.

12

Pablo's Surprise

The impact of the genocide reached far beyond Rwanda. Some 2 million people fled when the killing began. Many of them poured across the border to eastern Congo.

On one of several visits to Rwanda in the years following the genocide, Amy drove through a refugee camp and was surprised by the calm appearance of the people. They had been in the camp for two years and seen so much suffering. Yet they seemed to be making the best of their not-so-temporary living conditions. Many milled around, talking to one another. Children played in the rows between makeshift tents.

Amy *(opposite)* helps a Rwandan student watch a black-cheeked mangabee. Education, she feels, is key to wildlife conservation and offers children experiences that help them appreciate native wildlife, such as the mountain gorilla above.

~ Familiar Territory

Amy was visiting Kahuzi-Biega National Park as a member of a WCS team conducting a survey of Grauer's gorillas. As she went about her work in the park, Amy remembered her initial visit to the area as a Peace Corps volunteer. Much had changed in the two decades since she first glimpsed wild gorillas here. Thousands of Kahuzi gorillas had been caught in the middle of a war. Gun-toting rebels killed some of them, trying to make a quick buck selling gorilla parts on the black market. No one knew how many were left. But a highland census showed a large decline. The park staff was devoted to protecting gorillas. Many had stayed to defend the park during the war, even when they stopped getting paid.

The region's political situation had not yet returned to normal, however. A few weeks into the survey, the local Congolese told Amy's team that trouble might be brewing: They had better finish the survey and leave the country, for a $1,000 bounty had supposedly been placed on American heads. Feeling vulnerable but not overly concerned (such rumors were common), Amy and three other American researchers sat around a campfire one night: "Wow, $1,000!" someone exclaimed. "They could get $4,000 right here!"

Amy and her fellow researchers take a break during their 1996 census of the gorilla population in Kahuzi-Biega National Park, Zaire.

Little did they know that the rumor had hit international newspapers. Back home, Amy's mother called Bill to ask, "Isn't Amy over there?" Bill assured Marion that Amy would be okay. Still, he was glad when she safely made it home a few weeks later.

~ *From Hiking Boots to Flowered Skirts*

In 1993, a year before the genocide in Rwanda, Amy took over Bill's position supervising the Wildlife Conservation Society's Africa programs. Her husband offered her some good advice: "Don't forget why we do this work. Be sure to make time to get out into the field." So Amy returned to Africa whenever she could, visiting WCS researchers who study everything from gorillas, elephants, and antelopes to baboons, chimps, and bush pigs. She also spent time meeting with government officials, which meant she could go from traipsing through the bush in muddy, sweat-soaked field clothes one day to holding high-powered discussions in a skirt, blouse, sandals, and earrings the next.

Noah and Ethan got used to seeing their mother leave for trips of more than a month. When Amy was home, Bill often left to travel for his new job as the director of WCS's North American programs. They didn't see each other as much as they wanted to, but Amy and Bill arranged their trips so that one of them was always home with the boys, helping with homework or coaching lacrosse and soccer.

When one parent was gone, the boys had the other to themselves. As a result, they developed strong bonds with each.

~ *Bringing the Jungle to the Bronx*

Because the Wildlife Conservation Society is based at New York City's Bronx Zoo, part of Amy's job was to work with zoo staffers to design exhibits that would teach zoo goers about wildlife and conservation. In the late 1990s, she helped the zoo create a one-of-a-kind exhibit aimed at showcasing the wonders of the African Congo forest.

Amy and the president of Congo lead a delegation celebrating the opening of the Bronx Zoo's Congo Gorilla Forest exhibit in June 1999.

Beneath an umbrella of leaves and branches, zoo visitors wind along an elephant footpath that cuts through the heart of a misty African rain forest. Each stop along the way is like a page from Amy Vedder's memory book—giant trees filled with colobus monkeys, a foot-long millipede like the ones Noah and Ethan used to play with, a creek bed surrounded by a big family of colorful mandrills.

Eventually, visitors find themselves in a darkened amphitheater. There they see a powerful eight-minute movie starring Amy. The film transports zoo visitors to the bush, where a herd of elephants feeds by a watering hole and a family of gorillas meanders through a dense forest of trees. As Amy and a Congolese researcher track the gorillas, they find spent shotgun shells next to a nest of leaves where a gorilla had slept the previous night—a discovery that makes the children in the audience gasp.

Residents of "Congo in the Bronx" include lowland gorillas *(above)* and other primates, colorful birds *(below)*, mandrills, red river hogs, amphibians, and reptiles.

But the movie ends on a note of hope for successful gorilla conservation: As the screen fades to black, a giant curtain slides open to reveal a real family of gorillas resting peacefully in a field of grass.

Much of the exhibit was designed to provide two big families of gorillas with a natural setting—giant trees with fallen branches that provide the perfect jungle gym for youngsters, hollow tree trunks that offer a dry spot during a rainstorm, and grassy clearings that furnish cozy resting places for mid-morning siestas.

Scattered throughout the Congo exhibit is a theme that has become Amy's motto: Discover, involve, protect. That is to say, discover the facts (study the wildlife), involve the local people, and protect the wildlife and habitat. At the end of the exhibit, visitors vote on how their $3 entry fee will be spent for wildlife conservation in Africa. Since the exhibit's creation in 1999, it has financed the salaries of African "eco-guards" who patrol protected forests and ward off poachers. Proceeds from the exhibit have also preserved a block of African forest so wild that the chimps, gorillas, and elephants living there have seen few, if any, humans.

~ A Greater Reach

In 1999 Amy took on a new post at the WCS. She now oversees projects that range from North America to Asia, spanning many areas where wildlife and people come in conflict. One of those projects is in the Greater Yellowstone ecosystem, where, at the age of 22 years, Noah joined a group of field researchers studying coyotes and pronghorn. (Ethan, 18, also showed an interest in science, attending college to become a science teacher.) Another conservation project is based in the Adirondacks, not far from where Amy and Bill own a family cabin, just as her parents did.

Amy also oversees conservation work in places she hasn't even visited yet. The projects aim to protect all sorts of wildlife, from the marine creatures living in a coral reef along the coast of Belize to jaguars, parrots, and monkeys in an Amazon rain forest, and from elephants, apes, and Nile crocodiles in Africa to gazelles in Mongolia. Still, Rwanda is never far from Amy's heart.

~ Gorilla Dynasty

Amy Vedder has returned to Rwanda many times over the years. In 2000 she went back to Mount Visoke just to check on an old friend. She found many familiar markers on the trail up the mountain—rock walls, fallen trees, boot-sucking mud—but she also noticed that the path was less shady than before. Many trees had been cut down during the genocide and its aftermath.

At Visoke's summit, Amy filled her lungs with the thin mountain air, breathing a little less easily than she wanted to admit and listening intently for the gorillas she knew were nearby. When she heard the crunch of a gorilla foot, her heart rate quickened. Tiptoeing through a dense thicket, she emerged on a dreamy scene: dozens of big, shaggy mountain gorillas feeding together on the ridge.

Happy ending? Pablo *(left)* grew up to lead the largest family of mountain gorillas ever recorded. Later, Cantsbee *(right)* would be the leader of Pablo's dynasty.

Amy scanned the crowd and spotted Pablo, the mischievous gorilla who had tried to adopt her as his mother 22 years earlier. Now, Pablo was a great silverback with a family of his own.

Calming any fears that Amy had harbored about the little ape being able to head his own brood, Pablo was the kingly leader of 44 gorillas—the biggest mountain gorilla family ever recorded.

One day Pablo, like his father before him, would yield to a younger, more vibrant ruler. In fact, by the time Amy returned for another visit in 2003 with Bill, Noah, and Ethan, the great Pablo had indeed been unseated by a younger silverback. The new leader was also one of Amy's extended family members—Cantsbee, the one that had tested five-year-old Noah's courage by trying to untie his shoelace. With Cantsbee as the leader of Pablo's dynasty, Group 5 had swelled to 51 gorillas!

Actually, the total mountain gorilla population in the Virungas had grown from 260 animals in 1978 to 380 in 2004. Amy was overjoyed that Pablo and the others had survived, beating all expectations, in the face of so much turmoil. And there was more: The Nyungwe Forest would soon become a national park, protected for the future.

Twenty-five years had passed since Amy and Bill first came to Rwanda to help save the mountain gorillas. The population that once seemed destined for extinction was now thriving, and the place that had once been threatened by development was protected for its unique wild inhabitants. *We—and a lot of dedicated Rwandans—really did make a difference,* Amy thought to herself. Then Amy Vedder turned reluctantly on her heel and walked back down the mountain.

Timeline of Amy Vedder's Life

1951	Amy Vedder is born on March 24.
1959	World-renowned ecologist George Schaller spends a year studying the mountain gorilla, deflating the myth that gorillas are monsters like King Kong and estimating that the Virunga gorilla population contains 400 to 500 animals.
1967	Dian Fossey goes to Africa to study gorillas.
1969	Amy begins college at Swarthmore in Pennsylvania and meets future husband Bill Weber during freshman orientation.
1972	Amy marries Bill on the lawn at Swarthmore College on June 3.
1973	Amy graduates with a bachelor's degree in biology. Soon after, she and and Bill join the Peace Corps and are sent to central Africa. Researchers determine the Virunga mountain gorilla population has dropped to between 250 and 275 animals and could soon become extinct.
1975	Amy and Bill return to the United States, determined to find a graduate school where Amy can study gorilla ecology and Bill can make the leap from political science to wildlife conservation. They begin classes at the University of Wisconsin in Madison.
1977	Digit, a mountain gorilla made famous by gorilla researcher Dian Fossey, is killed in Rwanda.
1978	Amy and Bill arrive at Dian Fossey's Karisoke field station. Weeks later, they rescue a gorilla illegally taken from Volcanoes National Park and name her Mweza. Group 4 gorillas are massacred by poachers. Bill completes a gorilla census, determining that 260 mountain gorillas still survive in Rwanda. The population is stable, with many young.
1979	The Mountain Gorilla Project officially gets off the ground.
1980	Noah Gerhardt Weber is born on November 1.
1982	Amy earns a master's degree in zoology from the University of Wisconsin.
1983	Amy returns to Africa for four month's work in Burundi.

1985	Ethan Heller Vedder is born in Madison, Wisconsin, on March 1. Amy returns to Rwanda with Ethan and Noah, joined by Bill. Dian Fossey is murdered in her cabin at Karisoke on December 26. Amy begins biological surveys and field research in the Nyungwe Forest.
1987	Bill is offered the position of deputy director of the Wildlife Conservation Society's (WCS) international program. Amy stays in Rwanda to finish her fieldwork in the Nyungwe Forest.
1988	Amy is reunited with her family and moves into their new home in New York. Amy begins teaching ecology at Fordham University in September.
1989	Amy earns a Ph.D. in zoology from the University of Wisconsin. She becomes the Biodiversity Program Coordinator for WCS.
1990	Political uprising begins in Rwanda.
1993	Amy is promoted to director of WCS's Africa Program.
1994	Rwandan genocide begins. During a three-month killing spree, 800,000 Tutsi are murdered.
1995	Amy returns to Rwanda to find that many colleagues and friends disappeared or were murdered in the genocide. The country's population hits a low of 4.8 million.
1996	Amy takes part in the Kahuzi-Biega National Park gorilla census in the Democratic Republic of Congo.
1999	The Congo Gorilla Forest exhibit opens at the Bronx Zoo. Amy becomes director of WCS's Living Landscapes program; begins overseeing conservation projects all over the world.
2000	Amy visits Pablo, now the leader of the biggest family of mountain gorillas ever recorded.
2003	Amy visits mountain gorillas again with Bill, Noah, and Ethan. Pablo's family has grown even larger.
2005	Nyungwe Forest becomes the newest national park in the world. Amy works for conservation in Asia and Latin America, as well as Africa and the United States.

About the Author

Rene Ebersole writes about all things wild for a wide range of audiences. Her articles on science, nature, and the environment have appeared in such publications as *National Wildlife, Audubon, Wildlife Conservation, Current Science,* and *National Geographic Explorer.* She thinks science writing is one of the coolest jobs on the planet because each story is an adventure and an opportunity to meet an inspiring person who is making a difference. She and her husband Michael live near New York City, where Rene works as a senior editor for *Audubon* magazine.

GLOSSARY

This book is about a biologist who studies primates. To figure out the meaning of scientific words, it helps to know a little Greek and Latin. Let's start with *biology*. The word comes from the Greek *bios* meaning "life" and the Latin *logia* meaning "science." A *biologist* is a scientist who studies living organisms.

The word *primate*, comes from the Latin word *primus*, meaning "first" or "leader." A primate is any of the order of mammals that include apes, monkeys, and us—humans. So *primatology* is the study of primates, especially those other than recent humans, and a *primatologist* is someone who studies primates.

Here are some other science words you will find in this book. For more information about them, consult your dictionary.

canopy: the umbrella-like upper level of branches and foliage in a rain forest

census: an official count of a population in a region

condition: to shape the behavior of an animal by repeated exposure to a specific set of conditions or situations

CPR (cardiopulmonary resuscitation): a procedure to help someone whose breathing and heartbeat have stopped. From the Greek *cardia* meaning "heart" and the Latin *pulmonarius* meaning "lung."

dung: waste matter from the intestines of animals; feces.

ecology: the study of the relationship among plants, animals, and other living things and their environment. From the Greek *oik*, meaning "environment."

entomology: the study of insects. From the Greek *entomon* meaning "insect."

ethnic cleansing: the execution of people because of their race or religious beliefs; also known as genocide.

extinct: the death or disappearance of a plant or animal species; no longer existing. From the Latin *exstinctum* meaning "extinguished."

habitat: the natural environment of an animal or plant

hypothesis: a possible explanation or theory that can be tested by observation or experiment

machete: a broad heavy knife used as a tool and weapon

noseprint: the pattern of deep lines that crease the skin above a gorilla's nose. Gorilla researchers use these distinct patterns to identify individual animals.

nutrients: food substances such as proteins, carbohydrates, vitamins, and minerals needed for good health. From the Latin *nutrient* meaning "nourish."

ornithology: the study of birds, including their body structure, life cycle, and behavior

poachers: illegal hunters of animals

silverback: a mature male gorilla, identifiable by the silver hair on his back

snare: a trap consisting of a noose or set of nooses for catching animals

species: a group of living things that are the same in many ways and are able to mate and produce offspring

subspecies: a group within a species that has recognizable and distinct characteristics

topographic map: a map that shows in detail the features of a landscape. From the Greek *topos* meaning "place."

zoology: a branch of biology that focuses on the study of animals

Metric Conversion Chart

When you know:	Multiply by:	To convert to:
Inches	2.54	Centimeters
Feet	0.30	Meters
Yards	0.91	Meters
Miles	1.61	Kilometers
Acres	0.40	Hectares
Pounds	0.45	Kilograms
Gallons	3.79	Liters
Centimeters	0.39	Inches
Meters	3.28	Feet
Meters	1.09	Yards
Kilometers	0.62	Miles
Hectares	2.47	Acres
Kilograms	2.2	Pounds
Liters	0.26	Gallons

Further Resources

Women's Adventures in Science on the Web

Now that you've met Amy Vedder and learned all about her work, are you wondering what it would be like to be a wildlife biologist? How about an astronomer, a forensic anthropologist, or a robot designer? It's easy to find out. Just visit the *Women's Adventures in Science* Web site at **www.iWASwondering.org.** There you can live your own exciting science adventure. Play games, enjoy comics, and practice being a scientist. While you're having fun, you'll also get to meet amazing women scientists who are changing our world.

BOOKS

Bürgel, Paul Hermann and Manfred Hartwig. *Gorillas.* Minneapolis: Carolrhoda Books, 1992. Full-color photographs and a lively text take the reader on a tour through the habitat of the mountain gorillas of the Virungas.

Simon, Seymour. *Gorillas.* New York: Harper Trophy, 2003. This illustrated book offers information about three of the four gorilla subspecies, including their physical differences, habitats, diets, and behaviors.

Stewart, Kelly J. *Gorillas: Natural History & Conservation.* Stillwater, MN: Voyageur Press, 2003. This book offers a color essay written by a research assistant of the late Dian Fossey.

Weber, Bill and Amy Vedder. *In the Kingdom of Gorillas: Fragile Species in a Dangerous Land.* New York: Simon and Schuster, 2001. This autobiography details the couple's adventures in Rwanda as they worked to save the country's mountain gorillas from extinction.

WEB SITES

African Wildlife Foundation Gorilla Page: http://www.awf.org/wildlives/149
Keep up-to-date on recent news and regional conservation efforts to save the habitats of mountain gorillas in Rwanda, Uganda, and Congo.

Dian Fossey Gorilla Fund International: http://www.gorillafund.org/
Learn more about the ongoing activities and field research of DFGFI founded by Dian Fossey in 1978 and dedicated to the conservation of gorillas and their habitats in Africa.

Exploring the Environment of Mountain Gorillas:
http://www.cotf.edu/ete/modules/mgorilla/mgbiology.html
Explore the social behavior, habitat, and diet of Africa's mountain gorillas at NASA's Classroom of the Future site.

Wildlife Conservation Society Gorilla Page:
http://www.congogorillaforest.com/congoconservationchoices/congogorillaconservation
Take a virtual tour and meet the residents of the Congo Gorilla Forest at the Bronx Zoo.

World Wildlife Fund Gorilla Page:
http://www.worldwildlife.org/gorillas/subspecies.cfm
Find out how and why the four subspecies of African mountain gorilla are making a comeback.

Selected Bibliography

In addition to interviews with Amy Vedder, her family, friends, and colleagues, the author did extensive reading and research to write this book. Here are some of the sources she consulted.

Fossey, Dian. *Gorillas in the Mist.* Boston: Mariner Books, 2000.

Nichols, Michael, George B. Schaller, and Nan Richardson. *Gorilla: Struggle for Survival in the Virungas.* New York: Aperture, 1992.

Schaller, George B. *The Year of the Gorilla.* Chicago: University of Chicago Press, 1988.

Weber, Bill and Amy Vedder. *In the Kingdom of Gorillas: Fragile Species in a Dangerous Land.* New York: Simon and Schuster, 2001.

INDEX

A
Adirondack mountains, 4, 5–6, 101
Amin, Idi, 12, 13
Akagera National Park, 81, 92
Austen, Kathleen, 73–74

B
Baboons, 70
Belize, 101
Benda Lema, 61
Beringe, Oscar von, 24–25
Blue monkeys, 80
British Flora and Fauna Preservation Society, 53
Bronx Zoo, 19, 88
  Congo Gorilla Forest exhibit, 99–100
Burundi, 68–69, 92
Bush babies, 70
Bwindi Impenetrable Forest, 66–67

C
Cameroon, 22, 23
Cape buffalo, 46
Caroga Lake, 5–6
Chimpanzees, 70, 81–82
Colobus monkeys, 70, 83, 100
Cornell University, 18
Coyotes, 101
Cross River gorilla (*Gorilla gorilla diehli*), 22, 23
Crowned eagles, 70

D
Democratic Republic of the Congo, 11, 22, 92. *See also* Zaire
deSchryver, Adrien, 17

E
Ecotourism, 57, 58, 59–61, 62–63, 65
Earthworms, 80
Endangered and threatened species, 23, 24

F
Fordham University, 89
Fossey, Dian
  Amy's and Bill's relationship with, 38, 49, 74, 75–76
  burial, 76
  field station, 25
  Gorillas in the Mist book, 18
  and Mweza's injury and death, 30, 32, 33, 34, 38, 74
  murder, 72–77
  personality, 75–76
  primate research and conservation, 18–19, 51
  staff relationships, 17, 34, 74

G
Gahinga mountain, 23, 67
Gabon, 22
Gorillas. *See also* Lowland gorillas; Mountain gorillas
  behavior, 25
  discovery, 24–25
  noseprint identification, 43
  species, 22–23
  studies of, *see* Primatology
  taboos on eating, 29
Greater Yellowstone ecosystem, 101

H
Hagenia trees, 25, 70
Habyarimana, Juvenal, 93

J
Jonathan Dickinson State Park, 9–10

K
Kahuzi-Biega National Park, 16–17, 97–98
Karisimbi mountain, 22, 25, 67
Karisoke Research Center, 17, 18, 19, 25, 26–27, 29, 37, 46, 49, 76
Kennedy, John F., 9
King Kong myth, 25

L
Lake Kivu, 13
Leakey, Louis, 18
Loiselle, Bette, 68–69
Lowland gorillas
appearance, 22–23
Casimir, 16–17
eastern (*Gorilla beringei graueri*), 22, 24, 97–98
populations, 22–23, 97–98
western (*Gorilla gorilla gorilla*), 22, 24

M
Mangabeys, 80
Mikeno mountain, 33
Millipedes, 100
Mandrills, 100
Mobutu Sese Seko, 13, 32
Mona monkeys, 80
Mountain Gorilla Preservation Fund, 53
Mountain Gorilla Project, 45, 53, 54, 58–61, 65
Mountain gorillas (*Gorilla beringei beringei*)
aggressive behavior, 39–41, 60, 62–63
appearance, 23
Beethoven, 2–3, 29, 39, 41, 42, 47, 57
Brutus, 59–60, 62, 68
Cantsbee, 82, 102–103
conservation efforts, 45, 53–55, 57, 58–61, 62–63, 65
Digit, 18, 19, 21, 24, 27, 49, 76
Effie, 42, 47
family relationships, 1–3, 42
fieldwork with, 36–47
foods and feeding behavior, 40–42, 44, 45, 47, 49, 52
friendly behavior, 1–3, 18, 27, 29, 31, 34, 42–43, 47, 63, 82
graveyard, 76
Group 4, 18, 49–51
Group 5, 29, 37–47, 57, 82, 102–103
Group 6, 59–60
Group 11, 62
habitats, 23, 49, 51–52, 66–67, 101–102
home range, 40
Icarus, 42
Kweli, 27, 29, 50, 51
Liza, 42
Macho, 50, 51
Marchessa, 42
Muraha, 42
Mweza, 30–35, 37, 38, 74
Pablo, 1–3, 42, 57, 82, 102–103
Pantsy, 42
Peanuts, 18
poacher attacks, 18, 19, 21, 27, 30, 31, 48–51, 53, 77
Poppy, 42
populations, 23–24, 37–38, 51, 103
protected reserves, 23
Quince, 42
Puck, 42
Rwandan civil war and, 95, 97–98, 101–102
Shinda, 42
silverback males, 2–3, 21, 29, 42, 51, 60, 62–63
snare injuries, 29, 30, 32, 34–35
Stilgar, 62–63
tracking, 27, 28, 38–39, 44, 100
Tuck, 42, 43
Uncle Bert, 48–51
vocal communication, 31, 39–40, 62–63
Ziz, 42, 47
Muhavura mountain, 66, 67
Murangira, Emmanuel, 94–95

N
National Park Service (U.S.), 68
Nigeria, 23
Nyungwe Forest, 64, 69–71, 73, 80–81, 83–85, 86, 88, 103

P
Peace Corps, 11–17, 21, 52

Peace protests, 9
Poachers, 18, 19, 21, 27, 30, 31, 48–51, 53, 77
Primatology
- data analysis, 44–45, 58
- Dian Fossey's contribution, 18–19, 51
- hypothesis, 44
- note taking, 40
- observing animals, 36–47
- publishing results, 45
- tracking animals, 27, 28, 38–39
- transcribing notes, 44

Pronghorn antelope, 101

R

Rain forests, 2, 3, 64, 69–70, 101
Red-tailed monkeys, 80
Rwanda
- Akagera National Park, 81, 92
- Belgian rule, 92
- cattle-raising project, 51–52
- civil war and ethnic cleansing, 90–95, 97
- ecotourism, 57, 58, 59–61, 62–63, 65
- gorilla population, 1, 23, 24–25, 29
- Hutu, 91–92, 93, 94
- Murambi school massacre, 91, 94–95
- Nyungwe Forest, 64, 69–71, 73, 80–81, 83–85, 86, 88, 103
- refugee camps, 97
- Ruhengeri, 25
- political situation, 11, 12
- Tutsi, 91–92, 93, 94
- Volcanoes National Park, 1, 17, 19, 20, 21, 22–23, 30, 45, 51–52, 60, 61, 65, 66–67
- wildlife conservation, 53

Rwandan Patriotic Front, 92–93
Rwelekana, 29, 30, 38, 78

S

Sabyinyo mountain, 23, 66, 67, 81
Schaller, George, 25
Strangler fig, 80–81
Sunbirds, 70
Swarthmore College, 9–10

T

Turacos, 70

U

Uganda, 11–12, 23, 24, 66, 92
- Bwindi Impenetrable Forest, 66–67

University of Wisconsin at Madison, 17, 44–45, 52, 58, 67–68, 88–89
U.S. Agency for International Development, 68
Uwimana, Clementine, 79, 80, 94

V

van de Weghe, Gael, 80, 94
Vedder, Amy
- accidents and close calls, 5–6, 62–63
- Adirondack vacations, 4, 5–6
- captive in Uganda, 12
- in college, 7, 8, 9–10
- conservation dreams and accomplishments, 24, 45, 51–55, 57, 58, 97, 98–99, 100–101, 103
- children and family life in Africa, 65, 68–69, 71, 73, 79–83, 86–89
- Dian Fossey's relationship with, 38, 49, 75–76
- as ecotourist guide, 57, 62–63
- in elementary school, 7–8
- exploring Africa, 16–17
- gorilla research in Rwanda, 1–3, 16–17, 18, 19, 25, 27, 29–47, 57–58
- graduate school and dissertation, 17, 58, 65, 67–68, 88–89
- Group 5 fieldwork, 36–47
- growing up in Palantine Bridge, New York, 5–8, 46
- hiking the volcanoes, 66–67

Grauer's gorilla survey, 97–98
at Karisoke Research Center, 19, 25, 26–27, 46
life in Rwanda, 4, 44, 45, 46, 92–93
love of animals, 2, 5, 17
marriage and married life, 9–10, 11–17, 19, 24, 25, 26–27, 46, 58
Nyungwe Forest survey, 64, 69–71, 73, 80–81, 83–85, 86, 88
as Peace Corps volunteer, 11–15, 17, 21, 97
pets, 6, 7
rescuing hurt animals, 30–35, 50–51
Ruhengeri house, 79, 80, 93
science and nature interests, 7
teaching experiences, 14–16, 89
Wildlife Conservation Society career, 89, 97, 98–99, 101
Vedder, Barbara, 5, 6
Vedder, Chuck, 6, 7
Vedder, Ethan Heller, 71, 73, 78, 79, 81, 82, 86–88, 100, 101, 102
Vedder, Marilyn, 5, 6
Vedder, Marion, 7, 13, 98
Vedder, Nancy, 5, 6
Vervet monkeys, 70
Vietnam War, 9
Virunga volcanoes, 20, 21
Visoke mountain, 21, 22–23, 25, 45, 49, 67
Volcanoes National Park, 1, 17, 19, 20, 21, 22–23, 30, 45, 51–52, 60, 61, 65, 66–67

W

Weber, Bill
accidents and close calls, 56, 58–60, 68, 82
in Burundi, 68–69
children and family life in Africa, 79, 81–82, 88, 102
as ecotourist guide, 58, 59–60
gorilla census, 24, 38–39, 46, 49, 51, 52–53, 60
graduate school, 17, 19, 52
land-use management, 52, 68–69, 86, 89
life in Rwanda, 25, 26, 46, 61–62, 92–93
marriage, 9–10
as Peace Corps teacher, 11–15, 52
saving Kweli, 29, 30, 32–35
Wildlife Conservation Society career, 86, 88, 89, 99
as wildlife consultant, 68
wildlife education campaign, 53
Weber, Noah Gerhardt, 65, 68–69, 73, 79, 80, 81–82, 86–88, 92, 94, 100, 101, 102–103
Western Albertine Rift Mountains, 22
Wildlife conservation
ecotourism, 54–55, 57, 58, 59–61, 62–63, 65
education campaigns, 53, 54, 99–100
fundraising organizations, 19, 53, 71, 86
land-use management, 52, 68–69, 86, 89
pioneers, 52
preventing poaching, 65, 101
zoo exhibit, 99–100
Wildlife Conservation Society, 19, 71, 86, 88, 89, 97, 98–100
Wrangham, Richard, 17, 19

Z

Zaire
gorilla population, 22, 30–31, 32
Kahuzi-Biega National Park, 16–17, 97–98
map, 11
students in, 14–16
women's life in, 15–16

Women's Adventures in Science Advisory Board

A group of outstanding individuals volunteered their time and expertise to help the Joseph Henry Press conceptualize and develop the *Women's Adventures in Science* series. We thank the following people for their generous contribution of time and talent:

Maxine Singer, Advisory Board Chair: Biochemist and former President of the Carnegie Institution of Washington

Sara Lee Schupf: Namesake for the Sara Lee Corporation and dedicated advocate of women in science

Bruce Alberts: Cell and molecular biologist, President of the National Academy of Sciences and Chair of the National Research Council, 1993–2005

May Berenbaum: Entomologist and Head of the Department of Entomology at the University of Illinois, Urbana-Champaign

Rita R. Colwell: Microbiologist, Distinguished University Professor at the University of Maryland and Johns Hopkins University, and former Director of the National Science Foundation

Krishna Foster: Assistant Professor in the Department of Chemistry and Biochemistry at California State University, Los Angeles

Alan J. Friedman: Physicist and Director of the New York Hall of Science

Toby Horn: Biologist and Co-director of the Carnegie Academy for Science Education at the Carnegie Institution of Washington

Shirley Jackson: Physicist and President of Rensselaer Polytechnic Institute

Jane Butler Kahle: Biologist and Condit Professor of Science Education at Miami University, Oxford, Ohio

Barb Langridge: Howard County Library Children's Specialist and WBAL-TV children's book critic

Jane Lubchenco: Professor of Marine Biology and Zoology at Oregon State University and President of the International Council for Science

Prema Mathai-Davis: Former CEO of the YWCA of the U.S.A.

Marcia McNutt: Geophysicist and CEO of Monterey Bay Aquarium Research Institute

Pat Scales: Director of Library Services at the South Carolina Governor's School for the Arts and Humanities

Susan Solomon: Atmospheric chemist and Senior Scientist at the Aeronomy Laboratory of the National Oceanic and Atmospheric Administration

Shirley Tilghman: Molecular biologist and President of Princeton University

Gerry Wheeler: Physicist and Executive Director of the National Science Teachers Association

Library Advisory Board

A number of school and public librarians from across the United States kindly reviewed sample designs and text, answered queries about the format of the books, and offered expert advice throughout the book development process. The Joseph Henry Press thanks the following people for their help:

Barry M. Bishop
Director of Library Information Services
Spring Branch Independent School District
Houston, Texas

Danita Eastman
Children's Book Evaluator
County of Los Angeles Public Library
Downey, California

Martha Edmundson
Library Services Coordinator
Denton Public Library
Denton, Texas

Darcy Fair
Children's Services Manager
Yardley-Makefield Branch
Bucks County Free Library
Yardley, Pennsylvania

Kathleen Hanley
School Media Specialist
Commack Road Elementary
Islip, New York

Amy Louttit Johnson
Library Program Specialist
State Library and Archives of Florida
Tallahassee, Florida

Mary Stanton
Juvenile Specialist
Office of Material Selection
Houston Public Library
Houston, Texas

Brenda G. Toole
Supervisor, Instructional Media Services
Panama City, Florida

Student Advisory Board

The Joseph Henry Press thanks students at the following schools and organizations for their help in critiquing and evaluating the concept for the book series. Their feedback about the design and storytelling was immensely influential in the development of this project.

The Agnes Irwin School, Rosemont, Pennsylvania
La Colina Junior High School, Santa Barbara, California
The Hockaday School, Dallas, Texas
Girl Scouts of Central Maryland, Junior Girl Scout Troop #545
Girl Scouts of Central Maryland, Junior Girl Scout Troop #212

Illustration Credits:

Except as noted, all photos courtesy Bill Weber and Amy Vedder.

Cover © Andrew L. Young; **4-10** Courtesy Amy Vedder; **18** Peter Veit/National Geographic Image Collection; **19** Robert I.M. Campbell/National Geographic Image Collection; **24** Will Mason; **43** *(r)* Amy Vedder; **45** *(t)* Amy Vedder; *(b, both) American Journal of Primatology,* © 1984. Reprinted with permission of Wiley-Liss, Inc., a subsidiary of John Wiley & Sons, Inc.; **72, 75** © Yann Arthus-Bertrand/Corbis; **76** Michael Nichols/National Geographic Image Collection; **99** D. DeMello © Wildlife Conservation Society; **100** *(t)* D. Shapiro © Wildlife Conservation Society; *(b)* © H. David Stein

Maps: © 2004 Map Resources

The border image used throughout the book is a detail of the tree *Symphonia globulifera* appearing in full on p. 84, right.

JHP Executive Editor: Stephen Mautner

Series Managing Editor: Terrell D. Smith

Designer: Francesca Moghari

Illustration research: Christine Hauser and Joan Mathys

Special contributors: Meredith DeSousa, Allan Fallow, Colin Groves, Mary Kalamaras, April Luehmann, Mary Beth Oelkers-Keegan, Corrina Ross, Anita Schwartz, Liz Williamson

Graphic design assistance: Michael Dudzik